Fermented Meats

Fermented Meats

Edited by

G. Campbell-Platt

and

P.E. Cook

Department of Food Science & Technology
University of Reading
Reading

SPRINGER-SCIENCE+BUSINESS MEDIA, B.V.

First edition 1995

© 1995 Springer Science+Business Media Dordrecht
Originally published by Blackie Academic & Professional in 1995

Typeset in 10/12 pt Times by Photoprint, Torquay

ISBN 978-0-7514-0201-8   ISBN 978-1-4615-2163-1 (eBook)
DOI 10.1007/978-1-4615-2163-1

A catalogue record for this book is available from the British Library

Library of Congress Catalog Card Number: 94–78423

Preface

Preservation by fermentation is one of the oldest food technologies, and yet it continues to play an important role in meat preservation in many parts of the world. These processes can be relatively simple, with minimal microbial involvement, or more complex, involving defined ingredients and starter cultures with controlled environmental conditions. Most meat fermentations rely on the use of salt as an ingredient, sometimes with the addition of nitrate, nitrite and spices. In some cases the meat may be smoked and, as with some cheese fermentations, fermented meats may be ripened by moulds and yeasts. The preservation of meats by fermentation depends on the interaction of a number of environmental and microbiological factors including the pH, water activity, redox potential and the presence of preservatives and a competitive microflora.

The subject of fermented meats is an important but relatively specialised area of microbiology and food technology. Few books have specifically addressed this subject and the topic has usually been dealt with in reviews and research papers with a significant proportion of these being published in languages other than English. As far as we are aware, this volume is the first to bring together a selection of key topics relating to the production of fermented meats and their chemical and microbiological properties. The book begins with a general chapter on the properties of meat. This has been included to assist those readers who may be unfamiliar with the physical, chemical and microbiological properties of meat as a substrate for fermentation. This then leads into more specialised chapters on the history and diversity of fermented meats, followed by chapters on the microbiology, chemistry and preservation aspects of this technology. Fermented meats are popular foods in many parts of the world and the range of products is diverse. The final chapter considers fermented meat production and consumption in the European Union, a topic which has received very little attention in the literature.

The increasing popularity of fermented meats means that the book should appeal to both food scientists, technologists and microbiologists, in the academic sector and those involved in the production of these important foods.

G.C.-P.
P.E.C.

Contributors

H. Blom
MATFORSK, Norwegian Food Research Institute, Osloveien 1, N-1430 Ås, Norway

G. Campbell-Platt
Department of Food Science & Technology, University of Reading, Whiteknights, PO Box 226, Reading RG6 2AP, UK

P.E. Cook
Department of Food Science & Technology, University of Reading, Whiteknights, PO Box 226, Reading RG6 2AP, UK

R.H. Dainty
MATFORSK, Norwegian Food Research Institute, Osloveien 1, N-1430 Ås, Norway

S. Fisher
Meat & Livestock Commission, PO Box 44, Winterhill House, Snowdon Drive, Milton Keynes MK6 1AX, UK

B. Jessen
Chr. Hansen's Laboratorium, 10–12 Bøge Allé, PO Box 407, DK-2970 Hørsholm, Denmark

L. Kröckel
Bundesanstalt für Fleischforschung, Institut für Mikrobiologie, Toxikologie und Histologie, E.-C. Baumann-Straße 20, 8650 Kulmbach, Germany

R. Lawrie
Department of Applied Biochemistry & Food Science, School of Agriculture, University of Nottingham, Sutton Bonington, Loughborough LE12 5RD, UK

L. Leistner
Institut für Mikrobiologie, Toxikologie und Histologie, Bundesanstalt für Fleischforschung, E.-C. Baumann-Straße 20, 8650 Kulmbach, Germany

M. Palmer
Meat & Livestock Commission, PO Box 44, Winterhill House, Snowdon Drive, Milton Keynes MK6 1AX, UK

J.J. Pestka   Department of Food Science & Human Nutrition, Michigan State University, East Lansing, Michigan 48824–1224, USA

P. Zeuthen   Department of Biotechnology, The Technical University of Denmark, Danmarks Tekniske Hojskole, DK–2800 Lyngby, Denmark

Contents

1 The structure, composition and preservation of meat

R. LAWRIE

1.1 Introduction

Meat is defined in the *Oxford English Dictionary* as 'the flesh of animals used as food, now chiefly butcher's meat, excluding fish and poultry'. Since the predominant portion of the edible flesh of animal carcasses consists of muscular tissue, meat can be conveniently regarded as the *post mortem* aspect of muscles. The principal attributes of eating quality in meat, for which consumers value the commodity, viz. colour, texture (including tenderness and juiciness) and flavour, thus depend upon the structure and chemistry of muscle. The main purpose of muscle is to effect movement of the animal within its environment and that of its component structures within its body. This is achieved by contraction of the muscles, whereby their ends and the members to which these ends are attached, are caused to approach one another. Since movement in animals thus signifies the operation of muscular tissue, the latter is clearly a predominant feature, not only of the domestic meat animals in UK (cattle, sheep and pigs), and of smaller mammals such as rabbit and hare, but also of horses, deer, antelope, buffalo, goats, camels, dogs, cats and rats and all other mammalian species, and of the many representatives of avian, piscine and reptilian species, which are eaten in various parts of the world, according to local custom and availability.

Although scientific knowledge of muscle has been mainly derived from studies of this tissue in beef, mutton/lamb and pork (the terms in English for the meat of cattle, sheep, and pigs, respectively), the findings apply generally since the basic structure of muscles is similar between species and classes of animals. There is no *intrinsic* characteristic which makes the meat of one species more suitable for preparing fermented products than any other, apart from a relatively high content of residual glycogen (e.g. in the muscles of the horse).

1.2 Structure

The musculature as a whole is composed of distinct anatomical units (muscles) of which the size, shape and attachments vary and reflect the

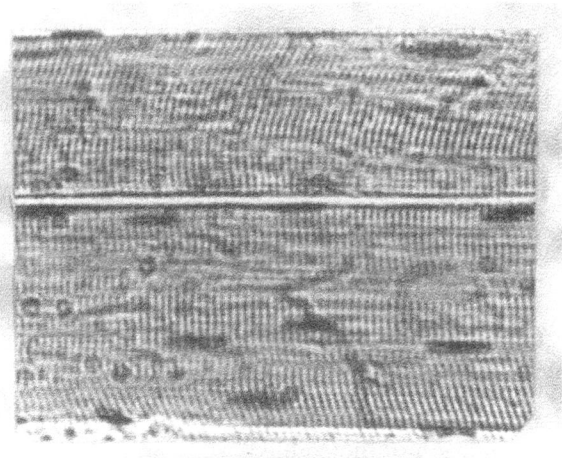

Figure 1.1 Two muscle fibres, showing cross-striations and nuclei (× 500).

kind of movement they have been evolved to effect. Despite this super-
ficial variation, all muscles have a similar structure, being composed of tens
of thousands of fibres which are aligned parallel to one another and to the
major axis of the muscle (Figure 1.1). These fibres are highly elongated,
multinucleated cells which, although only 10–100 μm, in diameter, may be
over 30 cm in length (Walls, 1960). Each fibre is bounded by a double
membrane, the sarcolemma. Within it there are large numbers of even
finer, parallel structures, the myofibrils (Figure 1.2). Under the light
microscope, fibres (and myofibrils) of skeletal muscles reveal cross-
striations i.e. a regular pattern of light and dark bands which run at right
angles to the long axis. Under the electron microscope each myofibril is
seen to be composed of large numbers of thick and thin filaments (Figure
1.3). The thick filaments are mainly aggregates of the molecules of the
protein myosin, the thin filaments mainly consist of a double helix of
globular molecules of the protein actin (Hanson and Huxley, 1953). The
dark (A) bands seen at low magnification are the regions where the thick
and thin filaments interdigitate; the light (I) bands are the regions which
contain only thin filaments. Under the electron microscope it is also seen
that the lengths of the myofibrils are divided at regular intervals by
transverse lines (Z-lines). The distance between two consecutive Z-lines
defines the smallest contractile unit of the muscle, the sarcomere. Sarcomeres comprise one-half of a light band, a dark band and one-half of the
next light band.

   Each myosin filament, which is a cylinder 1.5 μm in length and tapers
at each end, comprises some 200 individual myosin molecules which

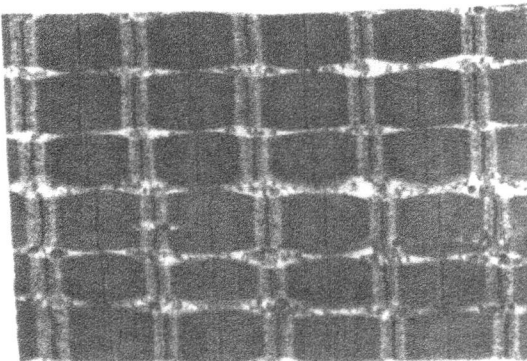

Figure 1.2 Electron micrograph of six myofibrils ($\times$ 10 000). The prominent lines which cross each myofibril at right angles to the long axis are the Z-lines. The distance between two adjacent Z-lines is the smallest unit of contractility, the sarcomere. Light zones (I-bands) on each side of the Z-lines consist mainly of actin filaments. In the dark zones (A-bands) of each sarcomere, thick filaments of myosin interdigitate with the thin actin filaments, except at the centre (the H-zone), where there are only myosin filaments, held together by proteins of the M-line (i.e. the less prominent lines at the midpoint of each sarcomere).

are themselves 150 nm in length and is surrounded by six actin filaments (Figure 1.3). Each myosin *molecule* has three sections: a 'tail' (L-meromyosin), a 'collar' which swivels (H-meromyosin, S2) and a 'head' (H-meromyosin, S1) (Huxley, 1963, 1971).

Located at intervals of 38.5 nm along the actin filaments there are assemblies of three other proteins, viz, troponins T, C and I (Schaub and Perry, 1969). The heads of the myosin, by extending laterally from the myosin filament, can attach to the adjacent actin filament via troponin C at these locations. During muscular contraction, the two ends of each sarcomere are caused to approach, as the heads of the myosin molecules link progressively more peripherally along the actin, the two sets of filaments interdigitating more and more, i.e. the muscle overall contracts. When the stimulus to contract ceases, the actomyosin links are broken and the sarcomeres extend to rest length (Figure 1.4).

The myofibrils within the fibre membrane are bathed in a fluid phase, the sarcoplasm, which contains soluble proteins, carbohydrates, fatty acids, salts and various insoluble organelles (lysosomes, peroxisomes and mitochondria).

Muscles are supported by a network of connective tissue, consisting mainly of the protein collagen, but with smaller amounts of elastin and minor glycoproteins and proteoglycans. The muscles exert their contractile power via these connective-tissue attachments to bones, muscles and other members. The connective tissue which surrounds the muscle as a whole is referred to as epimysium; septa of the latter penetrate into the body of the

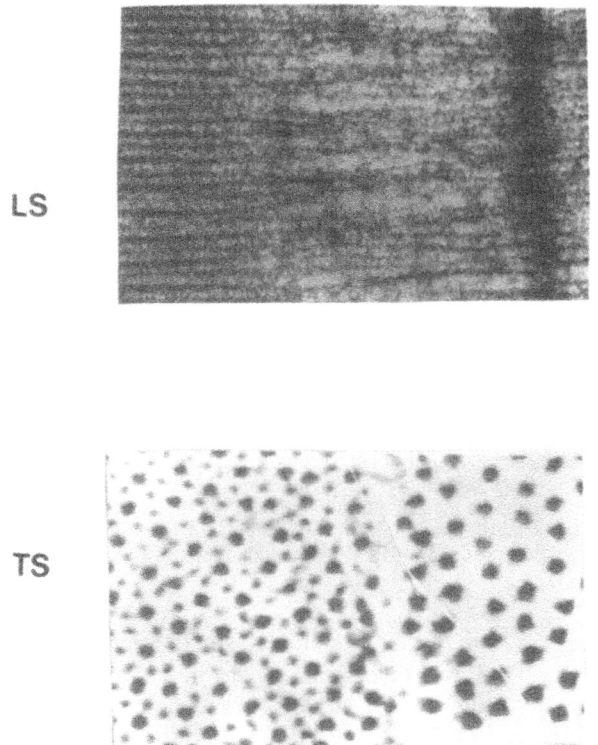

LS

TS

Figure 1.3 Electron micrographs of part of the central region of a sarcomere (× 200 000). LS: showing (from left to right) (i) region of interdigitating myosin and actin filaments, (ii) region consisting only of thick myosin filaments and (iii) the M-line. TS: showing, on left, each myosin filament surrounded by six actin filaments, where they interdigitate and, on right, myosin filaments only (in the H-zone). (Courtesy Dr H.E. Huxley, FRS.)

muscle, dividing it into groups of fibres (perimysium) and a layer of finer connective tissue surrounds each muscle fibre (endomysium) (Figure 1.5). Most collagen is itself composed of fibres, each of which, in turn, is built up from parallel (but quarter-staggered) aggregates of tropocollagen molecule which are 280 nm in length. Each tropocollagen molecule consists of three helical polypeptide chains, of which the amino acid composition varies but in each of which proline occurs at regular intervals (and accounts for the chemical properties of collagen) (Bailey and Light, 1989). In addition to these principal structural features, there is also a secondary 'cytoskeleton' in muscle, consisting of a number of other proteins, quantitatively small, including the recently characterised nebulin and titin of the so-called 'gap filaments' (Wang and Williamson, 1980; Locker 1987).

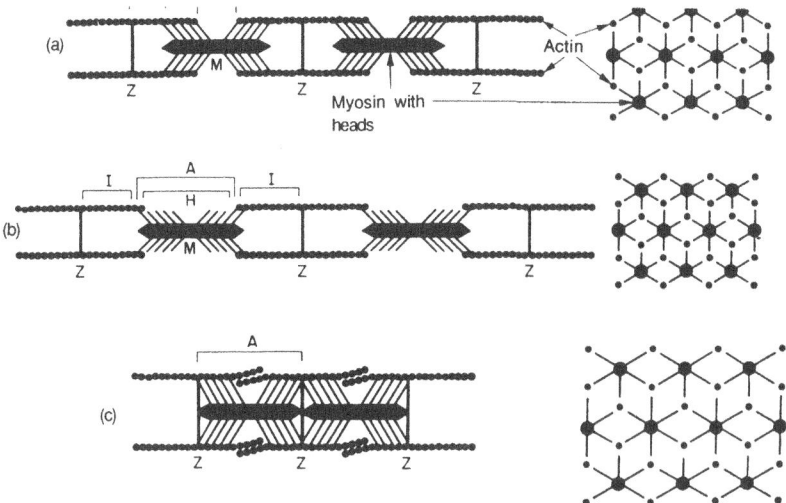

Figure 1.4 Schematic representation of two sarcomeres (a) at rest length (*ca.* 2.4 μm), (b) when stretched (*ca.* 3.1 μm) and (c) when contracted (*ca.* 1.5 μm). I = I (light) band: A = A (dark) band; Z = Z-line; M = M-line; H = H-zone. (Courtesy Dr J.R. Bendall and Pergamon.)

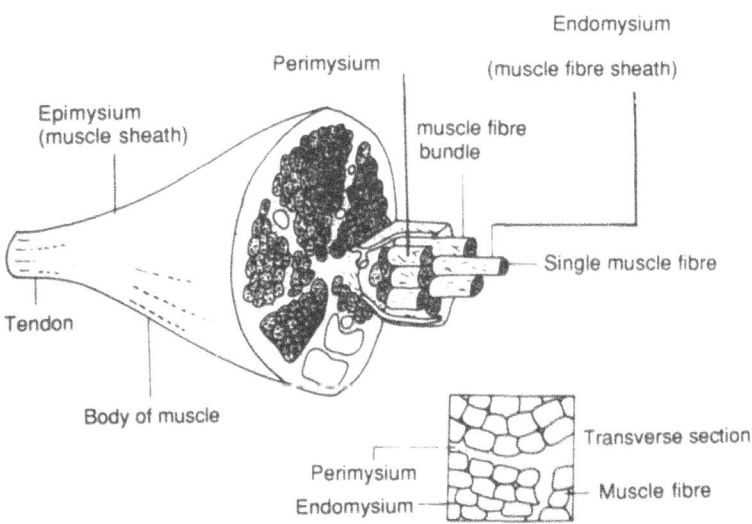

Figure 1.5 Diagram of muscle showing location of epimysial, perimysial and endomysial connective tissue. (Courtesy Prof A.J. Bailey and Elsevier

1.3   Chemical and biochemical nature of meat

1.3.1   Initial composition

In most instances, the composition of muscle will be assessed after the onset of *rigor mortis*, i.e. when the muscle has become inextensible. *Rigor mortis* is due to the irreversible union of the actin and myosin filaments where they overlap (the extent of which is determined by the amount of shortening which the muscles undergo immediately beforehand or concomitantly) (Bate-Smith and Bendall, 1949; Bendall, 1973). Just as during physiological contraction, such union is via the cross-bridges formed between the troponin complexes on the actin filaments and the heads of the myosin molecules on the ends of the myosin filaments. Actomyosin formation is inevitable when the level of adenosine triphosphate (ATP) falls below the level at which it can keep the muscle flexible. Even in muscle at rest, energy is required to maintain structural integrity and, in warm-blooded animals, to maintain temperature. The enzymic splitting of ATP to adenosine diphosphate (ADP) provides such energy, *In vivo* respiration very efficiently resynthesises ATP from ADP; but when the circulation ceases at death, the anaerobic metabolism which ensues is ineffective in this respect. The level of ATP can be temporarily maintained by its resynthesis from the reaction of ADP with creatine phosphate, of which there is a reserve of varying amounts in different muscles, and by the reactions involved in the conversion of glycogen to lactic acid (i.e. *post mortem* glycolysis); but the ATP eventually falls and *rigor mortis* occurs. The production of lactic acid causes the pH of the muscle to fall below the *in vivo* value (*ca.* 7.2) to an ultimate pH which depends on the amount of glycogen available in the muscle at the moment of death and, if the latter is sufficient, on the attainment of a value *ca.* 5.5, at which the enzymes effecting *post mortem* glycolysis are usually inactivated. These changes are shown in Figure 1.6.

The proximate composition of a typical, adult mammalian muscle *post mortem*, after the onset of *rigor mortis*, but before degradative changes commence, is 75% water, 19% protein, 2.5% lipids, 1.2% carbohydrate and 2.3% of miscellaneous non-protein substances (Table 1.1; Lawrie, 1991). Some 11.5% is myofibrillar protein (i.e. 60% of the total protein) consisting of the myosin and actin of the thick and thin filaments (together with minor quantities of other structural proteins), 5.5% is sarcoplasmic protein (i.e. proteins soluble in an aqueous medium, many being enzymes of the glycolytic pathway from glycogen *in vivo* to lactic acid *post mortem*), and 2.2% is contributed by the connective tissue and organelles. Of the lipids, most is neutral triglyceride; this is associated with phospholipids and fatty acids. Most of the 1% carbohydrate in *post mortem* muscle is lactic acid derived from glycogen in the living muscle. Of the *ca.* 2% of

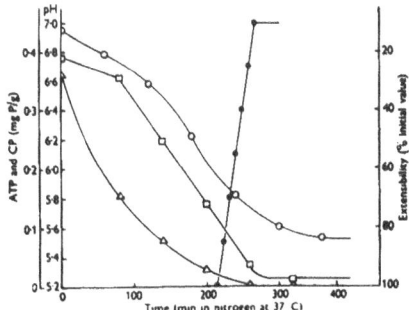

Figure 1.6 The onset of *rigor mortis* in *longissimus dorsi* muscle of the horse. Changes in ATP and creatine phosphate (CP) (as mg P/g) and in extensibility (as percentage initial value). □–□, ATP; △–△, CP: ○–○, pH; ●–●, extensibility. (From Lawrie, 1991; by permission of Pergamon Press Ltd.)

miscellaneous components, three-quarters are non-protein nitrogenous compounds, including amino acids and the remainder are various salts including those of potassium and phosphate. There are also traces of various vitamins, especially those of the B group.

A number of factors impose variation on the relative quantities of these components of muscle, reflecting differences in the type of contraction undertaken (Lawrie, 1991). The most obvious are species, breed, sex, age, anatomical location, plane of nutrition and exercise; these may be exemplified. Thus, the colour of adult porcine muscles (due to the muscle pigment, myoglobin), is usually much paler than that of corresponding muscles of the adult bovine or ovine; the lipids in pork generally contain a greater percentage of unsaturated fatty acids than those of the other two domestic species.

Some breeds of pig have genetic complements which cause the conversion of glycogen to lactic acid during *post mortem* glycolysis to proceed either very rapidly (e.g. Piétrain) or to a greater final acidity (i.e. a low ultimate pH, e.g. Hampshire) than those of the majority of breeds (e.g. Large White), whereby a pale, soft and exudative pork is produced (Monin and Sellier, 1985). Again, a recessive gene causes the meat of certain beef animals to have less connective tissue, and to be more tender, than that of the majority of cattle, from which the gene is absent (Lawrie *et al.*, 1964).

Sex tends to influence the quantity of intramuscular fat. Males have less than females under comparable conditions. On the other hand, the castrated members of both sexes have more intramuscular fat than corresponding entire animals.

Animal age has a profound effect on the composition of muscle. Almost all components increase in concentration as a percentage of wet weight as animals mature, although water itself diminishes. Thus, the myoglobin

FERMENTED MEATS

Table 1.1 Chemical composition of typical adult mammalian muscle after *rigor mortis* but before degradative changes *post mortem* (reproduced from Lawrie (1991) by kind permission of Pergamon Press Ltd.)

Components		Percentage wet weight
Water		75.0
Protein		19.0
Myofibrillar		11.5
myosin* (H- and L-meromyosins and several light-chain components associated with them)	5.5	
actin*	2.5	
connectin (titin)	0.9	
N_2 line protein (nebulin)	0.3	
tropomyosins	0.6	
troponins C, T and I	0.6	
α-, β- and γ-actinins	0.5	
myomesin (M-line protein), C-proteins	0.2	
desmin, filamin, F- and I-proteins, etc.	0.4	
Sarcoplasmic		5.5
glyceraldehyde phosphate dehydrogenase	1.2	
aldolase	0.6	
creatine kinase	0.5	
other glycolytic enzymes	2.2	
myoglobin	0.2	
haemoglobin and other unspecified extracellular proteins	0.6	
Connective-tissue and organelle		2.0
collagen	1.0	
elastin	0.05	
mitochondrial etc. (including cytochrome *c* and insoluble enzymes)	0.95	
Lipid		2.5
neutral lipid, phospholipids, fatty acids, fat-soluble substances	2.5	
Carbohydrate		1.2
lactic acid	0.90	
glucose-6-phosphate	0.15	
glycogen	0.10	
glucose, traces of other glycolytic intermediates	0.05	
Miscellaneous soluble non-protein substances		2.3
Nitrogenous		1.65
creatinine	0.55	
inosine monophosphate	0.30	
di- and tri-phosphopyridine nucleotides	0.10	
amino acids	0.35	
carnosine, anserine	0.35	
Inorganic		0.65
total soluble phosphorus	0.20	
potassium	0.35	
sodium	0.05	
magnesium	0.02	
calcium, zinc, trace metals	0.03	
Vitamins		
various fat- and water-soluble vitamins, quantitatively minute		

* Actin and myosin are combined as actomyosin in *post rigor* muscle.

content of muscle increases swiftly from birth to maturity, as the light red colour of veal, in comparison with the dark red of beef, attests. The increase in myoglobin reflects a need to service with oxygen the enzymes which effect respiration *in vivo*; their concentration concomitantly increases as animals grow older (Lawrie, 1953). The concentration of the contractile proteins also increases but that of the connective tissue decreases. Since a significant proportion of the toughness of meat is due to its connective tissue, this finding seems at variance with the generally greater toughness of the meat from mature animals. However, the anomaly may be explained by the fact that the collagen from older animals is more cross-linked and intractible than that from the immature animal. (Bailey *et al.*, 1974; Bailey and Light, 1989). There is a steady increase in the titre of total nitrogen, on a fat-free basis, with increase of animal age/weight — an important fact for the analyst (Analytical Methods Committee, 1991). Intramuscular fat also increases with age but its degree of unsaturation tends to diminish.

The most complex factor influencing the composition of muscle is its anatomical location for this determines the kind of contraction it has been evolved to perform. In broad terms, muscles can be classified as 'white' or 'red'. The former tend to operate in short, quick bursts of activity, with frequent periods of rest to recuperate; the latter operate relatively slowly over prolonged periods. 'White' muscles contain a high concentration of those enzymes which convert glycogen, the muscle fuel reserve, into lactic acid under relatively anaerobic conditions and their pale colour reflects the fact that they do not require a large content of myoglobin to store oxygen. 'Red' muscles, on the other hand, contain large concentrations of respiratory enzymes and have a high concentration of myoglobin to service them under aerobic conditions (Table 1.2). It must be acknowledged, however, that there is a continuous gradation of activity between the two extremes among the several hundred distinct muscles in an animal body and it can be appreciated that there will be a corresponding variability in their composition, which must be expected to affect their quality as meat *post mortem* (Table 1.3; Lawrie, 1991).

Prolonged exercise (training) is associated with an increase in the myoglobin concentration of muscles and in that of the respiratory enzymes which it services. There may also be an increase in the content of glycogen. On the other hand, severe exercise in the immediate preslaughter period — and indeed any circumstance producing stress in the animal — will deplete the *in vivo* level of glycogen and may lead to an ultimate pH level higher than the normal 5.5. A high ultimate pH has profound and undesirable effects on the quality of the meat, causing an unpleasant dark colour, sticky consistency and diminished flavour (Bernard, 1877; Howard and Lawrie, 1956; Tarrant, 1981).

A high plane of nutrition is generally reflected by an increase in the

Table 1.2 Some relative differences between 'white' and 'red' muscles (after Lawrie, 1978)

Parameter	'White'	'Red'	References
Myoglobin concentration	low	high	Gunther (1921)
Capacity for aerobic power	low	high	Lawrie (1952)
Respiratory enzymes	scarce	plentiful	Lawrie (1953)
Glycolytic enzymes	plentiful	scarce	Lawrie (1953)
			Scopes (1970)
Control of actomyosin ATPase	high	low	Gergely et al. (1965)
Capacity for protein synthesis	high	low	Citoler et al. (1966)
Calpain activity	high	low	Goll et al. (1971)
Actin-myosin filament overlap	little	extensive	Goldspink (1970)
3-Methylhistidine in myosin	high	low	Johnson and Perry (1970)
Myosin gelation temperature	32–44°C	42–54°C	Young et al. (1992)
Collagen			
concentration	high	low	Bailey (1974),
			Bailey et al. (1974)
reducible cross-links	plentiful	scarce	
Z-lines	narrow	broad	Gauthier (1969)
Fibre diameter	broad	narrow	George and Naik (1958)
Mitochondria	scarce	plentiful	Paul and Sperling (1952)

Table 1.3 Some chemical parameters of various muscles (reproduced from Lawrie (1991) by kind permission of Pergamon Press Ltd).

Muscle	Moisture (%)	Intramuscular fat (%)	Iodine no.	Total nitrogen (% fat free)	Hydroxyproline (μg/g)
Beef					
L. dorsi (lumbar)	76.51	0.56	54.2	3.54	520
L. dorsi (thoracic)	77.10	0.90	56.6	3.47	610
Psoas major	77.34	1.46	52.9	3.30	350
Rectus femoris	78.07	1.49	67.8	3.40	550
Triceps (lateral head)	77.23	0.73	62.2	3.45	1000
Superficial dig. flexor	78.67	0.40	81.4	3.27	1430
Sartorius	77.95	0.58	64.0	3.33	870
Extensor carpi radialis	74.83	0.60	68.1	3.29	1160
Pork					
L. dorsi (lumbar)	76.33	3.36	56.3	3.77	670
L. dorsi (thoracic)	76.94	3.26	55.5	3.69	527
Psoas major	77.98	1.66	62.8	3.58	426
Rectus femoris	78.46	0.99	71.5	3.41	795
Triceps (lateral head)	78.68	1.84	67.0	3.46	1680
Superficial dig. flexor	78.87	1.90	65.3	3.35	1890
Sartorius	78.71	0.87	–	3.41	850
Extensor carpi radialis	79.04	1.39	69.7	3.36	2470

concentration of intramuscular fat, especially of saturated fat. In the non-ruminant, the nature of the nutrients is also important since ingested unsaturated fats in the feed are laid down in the intramuscular fat whereby its tendency to undergo oxidative rancidity during meat storage is increased. In sheep and cattle, the rumen microorganisms hydrogenate incoming fatty acids. In young animals, a diet which is deficient in iron can cause anaemia and limits the synthesis of myoglobin; the meat will be pale, as in the flesh of milk-fed veal calves.

The occasional incidence of marked inter-animal variation in the composition of muscle, when all known factors have been controlled, indicates that some as yet unknown influences must also be involved.

1.3.2   Changes on storage

Although the composition of muscles is determined by, and varies because of, the factors outlined previously, the meat they become *post mortem* is subject to subsequent change to an extent which depends on the length, and on the circumstances pertaining during, the period elapsing until it is consumed. Three types of agent are involved — biochemical, chemical and microbiological.

Apart from the enzyme systems required for anaerobic glycolysis and respiration, muscles contain at least two groups of proteolytic enzymes which may operate before *post mortem* glycolysis has ceased. They cause the increase in tenderness and flavour which arises in 'aged' or 'conditioned' meat; but this is not because they effect extensive breakdown of the connective-tissue proteins to their constituent amino acids, as was once supposed, since there is no liberation of free hydroxyproline (Sharp, 1959). One group includes the calpains which have pH optima above 6 — but may operate below this value — and depend on calcium ions for their activity. They attack strategic links which are necessary for structural integrity in troponin C, Z-lines (desmin), connectin (gap filaments) and the M-line protein to which the tails of myosin molecules are attached at the centre of the sarcomeres (Penny, 1974; Murachi *et al.*, 1981; Dransfield, 1992). The second group includes the cathepsins which have pH optima below 6 and are normally located within organelles (lysosomes). They attack strategic links in troponins C and I, the cross-links of the non-helical telopeptides of collagen, the mucopolysaccharides of the ground substance in which collagen fibres are disposed and attack very slowly the contractile proteins, myosin and actin (and certain other structural proteins of the myofibril) (DeDuve and Beaufay, 1959; Penny and Dransfield, 1979; Ouali *et al.*, 1987).

The rates of proteolysis during conditioning depend generally on the temperature at which the meat is held. A tenderness increment which takes 14 days at 0°C, takes two days at 20°C and one day at 43°C. At a given temperature, the extent of proteolysis increases with time and varies with species, animal age and type of muscle. Thus, at chill temperature, the relative rates in beef, lamb, rabbit and pork have been shown to be 0.17, 0.21, 0.25 and 0.33% per day (Dransfield *et al.*, 1981). The rate is greater in lamb than in sheep, in vealers than in cull cows and in so-called 'white' than in 'red' muscles (Ouali and Talmant, 1990). The effect of muscle type, although tenfold lower than that of temperature, is threefold greater than that due to species.

A number of other minor biochemical changes occur during meat storage; these include the liberation of ammonia from nucleotides and the hydrolysis of fats whereby free fatty acids are liberated.

Various non-enzymic, chemical changes affect meat during storage. In the absence of effective energy provision *post mortem*, proteins denature and lose their *in vivo* capacity to bind water (Weber and Meyer, 1933; Scopes, 1964). Meat thus tends to exude fluid with increasing storage time, but the extent is strongly determined by the ultimate pH of the meat. A normal ultimate pH (*ca.* 5.5), being near the isoelectric point of the principal muscle proteins, causes significant water loss, whereas a high ultimate pH (e.g. >pH 6), which arises in muscles which have a depleted content of glycogen at the moment of death, is associated with minimal exudation (Empey, 1933). On the other hand, muscles which attain an ultimate pH lower than 5.5, or in which the rate of *post mortem* glycolysis is rapid, yield meat which is excessively watery (pale, soft and exudative or PSE; Briskey, 1964). Prolonged storage causes loss of exudated water by evaporation, and in turn, this enhances the denaturation of the proteins, including that of the muscle pigment, myoglobin, which becomes oxidised to unsightly brown or greyish metmyoglobin on exposed surfaces (Brooks, 1935). The oxidation of myoglobin accelerates the concomitant oxidation of lipids, although the latter also occurs independently. The more highly unsaturated fatty acids (especially those of phospholipids) are particularly liable to oxidise during storage. Moreover, although such oxidation progresses faster at relatively high temperatures, it still occurs at below freezing temperatures, the long period of storage which these permit being associated with slow but cumulative change (Dahl, 1958). The production of free fatty acids and carbonyls causes the development of off-flavour.

With increasing time of storage, the carbonyls from oxidized fatty acids and from carbohydrates, react with the amino groups of amino acids and proteins, the products polymerising to form dark pigments and causing bitter flavours to develop. These are the changes of the Maillard reaction, of which the rate increases with the pH of the meat and high moisture content (Lea and Hannan, 1950).

The changes in composition on storage caused by the intrinsic enzyme systems of the meat are expected and mainly desirable; those caused by chemical reactions are generally undesirable. However, it is those caused by microorganisms which have gained access to the meat and which are most serious for eating quality and may be dangerous to the health of the consumer. When muscles die, the release of water from proteins and the increased concentration of small molecules derived from the breakdown of proteins, fats and carbohydrates, provide a richly nutrient medium for the growth of microorganisms.

A wide range of microorganisms — bacteria, yeasts and moulds — will grow on meat, and, accordingly, the changes they cause vary greatly. The subject was reviewed by Brown (1982) and reassessed by Waites (1988). In seeking nourishment, microorganisms may liquefy the tissue by attacking connective tissue with collagenase and hyaluronidase enzymes. Free amino acids may be metabolised, liberating various gases such as hydrogen, ammonia, carbon dioxide, methylamine, hydrogen sulphide and mercaptans. Glycogen may be converted to acetic and butyric acids. A specific histidinase yields histamine which affects membrane permeability. Myoglobin may be oxidised to brown metmyoglobin, converted into green sulphmyoglobin or choleglobin or broken down completely to yellow or colourless bile pigments. Moreover, various microorganisms produce their own pigmentation and thus discolour the meat. If the numbers of microorganisms are sufficiently numerous their aggregates will appear as slime or as other unusual features. The production of toxins by pathogens is a particularly dangerous possibility. Since microbial alterations are the most undesirable of the results of *post mortem* storage, they merit detailed consideration.

1.4 Microbial spoilage

1.4.1 Sources of contamination

Meat may become infected by microorganisms either endogenously in the living animal or by subsequent *post mortem* contamination. Overt disease caused by *Bacillus anthracis*, *Mycobacterium tuberculosis* or *Brucella abortis* would lead to condemnation by veterinary inspectors and the meat would not reach consumers. However, there are several conditions which would not be readily detectable *in vivo*. These include infection by various types of *Salmonella* (*S. typhimurium* in lambs and bovines, *S. cholerae suis* in pigs), *Listeria monocytogenes*, *Campylobacter jejuni*, *E. coli* (0157:H7), *Giardia lamblia* and *Yersinia enterocolitica*. Such organisms have been involved in outbreaks of food poisoning (Dolman, 1957; Galbraith *et al.*, 1987; Grau, 1988).

Post mortem contamination of meat, however, is the major cause both of organoleptic spoilage and of food poisoning. The most immediate source is the animal's blood. Although, in healthy animals, this will be free of pathogens, provided the reticuloendothelial system is operating normally, bacteraemia may develop in the immediate pre and post-slaughter period through failure of the mucous lining of the intestinal tract to prevent invasion of organisms into the blood and other tissues. The gastrointestinal contents include a vast number of microorganisms, particularly bacteria. Other main sources of exogenous contamination are the feet, hides or skin (to which soil and faecal matter may be attached), the water, instruments and receptacles used in the slaughter hall, personnel and airborne material (Empey and Scott, 1939; Stolle, 1988; Sheridan, 1992). Fortunately the majority of the bacteria present are not viable at chill temperatures and the swift cooling of carcasses suppresses them. Of the microorganisms which are viable at 0°C to −1°C, it has been shown that 90% are *Achromobacter*, 7% *Micrococcus*, 3% *Flavobacterium* and *ca*. 1% *Pseudomonas* spp. These are probably harmless psychrotrophs, albeit that they adversely affect meat quality. Recently, it has been demonstrated that the composition of the *Pseudomonas* population on the surface of beef, lamb and pork is mainly determined by the local environmental microflora at the abattoir at which slaughter takes place (Shaw and Latty, 1984). Representatives of the mould genera *Penicillium*, *Cladosporium*, *Alternaria*, *Sporotrichium* and *Thamnidium* are also present among the contaminants at the abattoir.

Contamination by infected personnel or healthy carriers is feasible at all stages after the death of the animal. The organisms concerned include *Salmonella* spp., *Shigella* spp., *E. coli*, *Bacillus proteus*, *Staphylococcus albus*, *Staph. aureus*, and *Clostridium perfringens*. These are pathogens, but they will not be a hazard to consumers unless temperatures have not been lowered sufficiently or food handling has been unhygienic. Certain pathogens, however, *can* grow at chill temperatures — viz. *Cl. botulinum* type E (usually overgrown by harmless psychrotrophs (Schmidt *et al.*, 1961)), *Listeria monocytogenes* (Mossel *et al.*, 1975) and *Campylobacter jejuni* (Grau, 1988). Their potential presence merits ongoing research to improve the efficacy of control measures. This involves the identification of hazards, understanding the risks they present, identifying the locations or stages at which control could be exercised (critical control points), selecting the options for such control (and for subsequent monitoring) and for implementing the degree of control, required — an approach referred to as 'hazard analysis: critical control points' (HACCP: International Commission on Microbial Specification for Foods, 1988). In a recent survey of sources of meat surface contamination and of its control in slaughterhouses, Stolle (1988) found that assays of Enterobacteriaceae and total aerobic counts were particularly indicative in applying the Hazard Analysis Critical Control Point (HACCP). Various procedures to restrict contamination of meat surfaces at the abattoir have been employed. Thus, with lamb carcasses, Smith and Graham (1978) found that a 10-second immersion in water at 80°C reduced contamination by coliforms by 99%. It is essential, of course, that surfaces are aseptically dried thereafter.

1.4.2   Modification through environmental factors

Apart from the availability of nutrients, the survival and growth of microorganisms in meat is determined by general factors — temperature, moisture availability, pH, the oxidation–reduction potential and the nature of the gaseous environment to which the organisms are exposed. Insofar as several of these determinants will be operating simultaneously, however, the behaviour of microorganisms is difficult to predict in any given environment

Microorganisms can be approximately classified by the temperature ranges at which they grow preferentially, as psychrotrophs (temperature optima, −2°C to 7°C), mesophils (temperature optima, 10–40°C) and thermophils (temperature optima, 43–66°C). In effectively chilled meat, the spoilage microflora will not normally be pathogenic and will consist predominantly of *Pseudomonas* spp. under aerobic conditions and of lactobacilli anaerobically. Anaerobic growth necessitates a greater degree of substrate breakdown to gain energy than aerobic conditions and yields products which are more unpleasant. With ineffective refrigeration in the immediate *post mortem* period, the deeper locations in the carcass may foster the growth of mesophils, if there is a reservoir of such organisms in the animal at death (e.g. in lymph nodes). Their growth may cause 'bone taint' and particularly offensive odours (Haines, 1937; Cosnett *et al.*, 1956).

Subtle differences due to the interaction of temperature and meat species are found with the prepackaged fresh commodity. Thus, whereas *Achromobacter* and *Pseudomonas fluorescens* are the main contaminants both at 3°C and 7°C in lamb and beef, lamb is spoiled by *Brochothrix thermosphaocta* at 5°C. Whereas beef is spoiled by Gram-negative bacteria (Barlow and Kitchell, 1966).

The nature of the microorganisms growing on packaged cured meats is also affected by the temperature of storage. At 5°C the microflora of packaged, sliced bacon consists mainly of a mixture of halophilic micrococci and lactic acid bacteria. At storage temperatures between 5 and 20°C, however, the lactic acid bacteria overgrow the micrococci; above 20°C, however, the lactic acid bacteria cease growing, whereas the micrococci now dominate, but belong to a different type (similar to *Staph. aureus* (Ingram, 1960)).

A reduction in the number of bacteria occurs when meat is held below the freezing point (i.e. *ca.* −1.5°C), but yeasts and moulds will grow down to $-5°C$. It is not the cold *per se*, however, which is responsible for inhibiting microbial growth, since some organisms will grow in supercooled liquid at $-20°C$, but rather the removal of available water as ice.

The availability of water is another major determinant of microbial growth; it is complementary to that of osmotic pressure. When the concentration of soluble, dialisable molecules such as salts and sugars, increases, microorganisms have increasing difficulty in obtaining sufficient water for their metabolism. The concept of 'water activity' (a_w) has proved useful in this context; a_w is the ratio of the water vapour of the medium to that of pure water at the same temperature (Scott, 1953, 1957). Bacteria grow from an a_w of just under 1.0 to 0.75; yeasts and moulds will grow slowly at an a_w of 0.62. Fresh meats, therefore, having an a_w of *ca.* 0.99 can spoil through the growth of a wide range of microorganisms. Food poisoning strains of *Staph. aureus* have an optimum a_w of 0.995 (Scott, 1953), but the rate of growth falls rapidly as the a_w is lowered.

Lowering the a_w by increasing the concentration of salt or sugar, as in curing, effectively prevents the growth of many organisms, but selects for the growth of salt-tolerant (halophilic) bacteria (Eddy *et al.*, 1960). Such bacteria frequently are capable of reducing nitrate and nitrite to produce nitric oxide whereby the pink pigment (nitrosomyoglobin) of cured meats develops.

It has long been appreciated that the pH of the medium markedly affects microbial growth, Most bacteria grow optimally at *ca.* pH 7 and little below pH 4 or above pH 9 (Cohen and Clark, 1919). Since the ultimate pH of meat will lie between the *in vivo* level of *ca.* 7 to *ca.* 5.5, it is evident that microbial growth on this commodity can be potentially vigorous. To reiterate, the amount of glycogen available in muscles at the moment of death will determine the amount of lactic acid produced during anaerobic *post mortem* glycolysis and thus the ultimate pH attained. The stress factors which lower such glycogen reserves — exhausting exercise, starvation, excitability of temperament, rough handling, inclement weather (Lawrie, 1966) — are important in determining the extent of microbial growth during storage. Moreover, in the absence of readily available carbohydrate, microorganisms attack amino acids causing early spoilage due to off-odours (Haines, 1937). Pigs are very susceptible to the reduction of muscle glycogen by even moderate stresses. A well-known example occurred during the restructuring of the pig industry in Northern Ireland in the 1930s, when the animals were slaughtered centrally at abattoirs instead of on the farms of origin. The stress of transport and of factory slaughter depleted glycogen reserves significantly, causing a high ultimate pH and spoilage of the cured meat under conditions when salt penetration into the meat was too slow to control the microbial population (Callow, 1937). A normal ultimate pH in cured meats is also desirable because it increases the concentration of nitrite which is in the form of undissociated nitrous acid. The concentration of the latter needed to control, for example, *Staph. aureus*, at an ultimate pH of 5.5 is only one-twentieth of that required at pH 6.9 (Castellani and Niven, 1955). In packaged fresh meats, a high ultimate pH fosters the growth of microorganisms which produce hydrogen sulphide and can lead to spoilage through the formation of green sulphmyoglobin (Shorthose *et al.*, 1972).

Another factor controlling microbial growth in meat is the oxidation–reduction potential (E_h) which is a function of oxygen tension and of the concentration of those molecules which are in their higher valency form. Immediately *post mortem* E_h is about $+150$ mV and the growth of anaerobes is inhibited; but the E_h falls swiftly to about -50 mV within a few hours when clostridia will be encouraged (Barnes and Ingram, 1956). Fortunately, provided the glycogen reserves have been sufficient, the lactic acid formed during *post mortem* glycolysis will then inhibit growth.

Oxygen is one of a number of atmospheric gases which affect microbial metabolism, and the concentration of oxygen which they will tolerate permits their division into aerobes, anaerobes and facultative anaerobes. The exposed surfaces of meat will support the growth of aerobes of the genera *Achromobacter* and *Pseudomonas*. Recent computer-assisted analysis has shown that 80% of the isolates from beef, lamb and pork were strains of *P. fragi* and 4% of *P. fluorescens* (Shaw and Latty, 1984). It will be evident that if meat is packaged — as it now is frequently — the atmosphere within the pack will vary according to such parameters as the permeability of the wrapping material to gases and moisture. In permeable wraps, the spoilage pattern tends to follow that on exposed surfaces; in impermeable wraps, residual respiratory activity of the muscles, and that of the microorganisms present, will produce carbon dioxide which suppresses the growth of aerobes. Lactobacilli and *B. thermosphacta* are relatively tolerant to carbon dioxide and they may prevail in vacuum-packed meat products. It has been found that 90% of the lactobacilli isolated from these products belong to one or other of two groups of streptobacteria, and include the recently characterised *Lactobacillus carnis* (Shaw and Harding, 1984).

Atmospheres of 100% carbon dioxide will inhibit the growth of both lactobacilli and enterobacteria (Gill and Penney, 1986). Since it is the psychrotrophic organisms which are inhibited by carbon dioxide, impermeable wraps (within which its concentration rises) are likely only to delay spoilage at chill temperatures. The modern practice of displaying packaged meats in strongly-illuminated cabinets may cause the temperature within packs to rise significantly above that of the cabinet itself through retention within the pack of absorbed radiant energy. Such gases as ozone and formaldehyde are lethal to microorganisms, but they tend to oxidise fats or taint the product otherwise. Moreover, since they are toxic, they are prohibited.

1.5 Modes of meat preservation

It will be evident that the requirements of microorganisms for survival and growth afford the possibility of their control by storing meat under conditions which fail to provide these requirements, or which directly inhibit microorganisms by agents inimicable to them.

1.5.1 Temperature control

Although it has been empirically appreciated from the earliest times that meat retained its eating quality better when kept cool, it was not until the invention of mechanical refrigeration 150 years ago that systematic temperature control was applied to the commodity. At about the same time, with the discovery of microorganisms by Pasteur, it was realised that the principal reason for the beneficial effects of early *post mortem* cooling and subsequent cold storage was the creation, thereby, of an environment in which the growth of most pathogens was largely prevented and that of non-pathogenic spoilage organisms greatly slowed. Although storage below the freezing point of meat (*ca.* −1.5°C) is much more effective in this respect, than refrigeration above it (i.e. chilling in the region of 0 to 5°C), it is normally associated with excessive exudation of fluid ('drip') on thawing and, at least as initially employed, caused other aspects of quality deterioration. For this reason, meat of intrinsically good quality was not usually frozen and was stored at chill temperatures. The most exacting requirements for chilled meat were those which had to be applied to ensure that beef exported to UK from Australia and New Zealand would still be attractive when offered for sale in London upwards of 6–8 weeks later. To achieve this, the strictest hygiene was essential at the abattoir in order to minimise the initial contamination, the carcasses or sides had to be cooled as quickly as possible and the chill storage temperature maintained within a narrow range (Scott and Vickery, 1939). Clearly the time needed to cool the meat from *in vivo* temperature will be determined not only by the bulk of the meat but also by the speed and temperature of the refrigerant air in the chiller. For example, it takes about 80 hours for the deep leg temperature of 260-kg beef sides to attain 10°C in air at 8°C travelling at 0.5 m/s, but, with 100-kg sides, only 16 hours in air at 0°C and a speed of 3 m/s (Bailey and Cox, 1976).

As refrigeration engineering has improved, the efficiency of chilling operations has markedly increased. It might be supposed that this development would have been an unqualified advantage, but, in the early 1960s, it became apparent that if the temperature of the muscles was reduced below *ca.* 10–15°C whilst they were still in an early prerigor condition (i.e. a pH > *ca.* 6), they tended to shorten (Locker and Hagyard, 1963). Meat which shortens during *post mortem* glycolysis and becomes fixed in *rigor mortis* in the shortened configuration, is tough when cooked. This is the phenomenon of 'cold-shortening'. Since its discovery, ways of overcoming it, without foregoing the advantages of rapid chilling, have become especially important. This is through the advent of centralised butchering and packaging, whereby consumer portion cuts may be prepared from the still warm carcass. Yet the extra manipulation involved in preparing such cuts makes them particularly liable to become contaminated by microorganisms; this increases the need for speedy refrigeration, and the potential danger of causing 'cold-shortening'. Although 'cold-shortening' could obviously be avoided by delaying speedy refrigeration until the pH had fallen to *ca*. 6, this would be undesirable both microbiologically and economically. If, however, the carcass or side is electrically stimulated immediately after slaughter, the rate of *post mortem* glycolysis is vastly accelerated and the pH falls to *ca*. 6 in about two minutes. Being no longer responsive to cold shock, the carcass or side may now be chilled as quickly as desired without the danger of 'cold-shortening' and toughening (Carse, 1973). The efficacy of electrical stimulation depends on the time after death at which it is applied, on the position of the electrodes, on the cross-sectional area of the meat, on the voltage (80–400 V) and amperage (2–5 A), the duration of its application (1–4 min), the frequency of the pulses and on the pulse width reviewed by Bendall (1980). The most useful combination is determined by economic factors and operating feasibility in the particular abattoir concerned.

Electrical stimulation, in addition to rendering muscles immune to 'cold-shortening' (and the toughening which the latter causes), also has a positive tenderising effect. This was known to Benjamin Franklin 200 years before the problem of 'cold-shortening' was encountered. Since electrical stimulation swiftly brings the pH of muscles to ≤ 6, at which the membranes of lysosomes are weakened, liberating the acidophilic, proteolytic enzymes they contain, the process is believed to promote the tenderising changes of conditioning although this explanation for the effect is not universally held (Dutson *et al.*, 1980; Takahashi *et al.*, 1984).

Insofar as the present tendency is the production of relatively small retail or consumer cuts from carcasses as soon as possible *post mortem*, it is evident that 'red' muscles could predominate in some and 'white' in others. Because the former are more susceptible to 'cold-shortening' than 'white' (Gergely *et al.*, 1965) they would appear to merit electrical stimulation preferentially; but, although this would protect them against 'cold-shortening' during subsequent rapid chilling, it could inhibit the conditioning enzymes, which, as already mentioned, are less prevalent in 'red' muscles. The muscles of pigs, being 'whiter', are less susceptible to 'cold-shortening' than those of cattle or sheep; but very rapid chilling *can* lead to toughening in pork, which can be alleviated by electrical stimulation (Gigiel and James, 1984). It will be evident that the circumstances necessary to ensure maximum tenderness in chilled meat and concomitant economy of operation require careful selection.

Provided abattoir operations have been hygienic and otherwise effective, chilled meat will deteriorate principally through surface changes. This is of particular importance when cuts having a high surface to volume ratio are the common mode of sale. There will be gradual loss of water and the desiccation will denature proteins, promoting the oxidation of myoglobin to brown metmyoglobin and synergistic oxidation of lipids (Watts, 1954). Evaporation can be minimised by increasing the humidity of the atmosphere, but the growth of psychrotrophs would be encouraged. Thus, during the storage of pork over 8 days at 1°C, the total titre of psychrotrophs and those of *Pseudomonas spp. Brochothrix thermosphacta* and Enterobacteriaceae were found to be low in meat of low ultimate pH (ca. 5.40), in comparison with those of meat of normal ultimate pH (ca. 5.58); they were higher in pork of high ultimate pH (ca. 6.41). The latter was the most susceptible to develop spoilage odours (Greer and Murray, 1988).

Now that a wide range of packaging materials is available, surface desiccation can be controlled; the relative permeability of the films to gases must be considered as well as that to water vapour at the particular temperature of storage envisaged. The permeability of packaging films to nitrogen is generally less than that to oxygen and the latter less than that to carbon dioxide, whereas permeability to water vapour, is much greater than to these gases (Paine and Paine, 1983). It will be evident that the interaction of these parameters will determine the pattern of the psychrophils which develops.

The efficacy of carbon dioxide in suppressing aerobic psychrophils — and in preventing discoloration due to traces of oxygen — has been exploited in exporting chilled lamb from New Zealand to distant markets. The lamb is packed in foil-laminate containers which are completely impermeable to oxygen, under 100% carbon dioxide. These packs are then placed within a master pack of foil-laminate also under 100% carbon dioxide. The appearance of the lamb does not begin to deteriorate until the master pack is opened when it then has a display life at chill temperature comparable to that of freshly prepared lamb (Gill, 1989). On exposure to the air, the purplish myoglobin changes to bright red oxymyoglobin. Although *Brochothrix thermosphacta* is tolerant of carbon dioxide, the organism will not grow if the pH of the meat is below 5.8, as it would be normally. The storage life is at least four months and substantially longer than that with vacuum packaging.

Various films permit vacuum packaging of meat. Meat of pH > 6, however, is unsuitable since the growth of certain anaerobes which produce hydrogen sulphide may be fostered, leading to the formation of green sulphmyoglobin.

Whereas chill temperatures delay spoilage, freezing can extend storage life more or less indefinitely, provided desiccation is avoided and the temperature is sufficiently low. Thus the flesh of mammals has been preserved for upwards of 15 000 years in the frozen tundra of Siberia. The major problem with freezing is the translocation of water which occurs when it forms ice and which is manifested as exudation (drip) on thawing (Chambers and Hale, 1932; Empey, 1933; Howard and Lawrie, 1956). The magnitude of the drip is determined by the speed of freezing, usually measured by the time taken for the temperature to fall from 0°C to −5°C. However, the effects of the rapidity of freezing differ, depending on whether the meat is subjected to freezing before or after the onset of *rigor mortis*. In the latter circumstance, the system is relatively static and it is the most usual. For post-rigor meat, the more rapidly it is frozen, the smaller the ice crystals, the less is the disruption of the cells, the less the denaturation of proteins, the greater the re-absorption of water on thawing and the less the drip (Cook *et al.*, 1926; Love and Haraldsson, 1961). With more or less instantaneous freezing, which is possible with minute portions of muscle, there is no drip on thawing. In practice, such rates are impossible, but especially with smaller cuts and with the low freezing temperatures and swift heat transfer now possible commercially, a useful reduction in drip can be achieved when the time to freeze is *ca.* 20 minutes in comparison with *ca.* 3 hours (Anon and Calvelho, 1980). With large portions of meat (e.g. sides of beef) the main advantage of rapid freezing, especially when applied to the warm carcass without any period of prior chilling, is the curtailment of evaporative losses. There is also some evidence, however, that insofar as protein denaturation is less with a slower rate of *post mortem* glycolysis, such rapid refrigeration diminishes drip.

Apart from the rate of freezing of post-rigor meat, a number of other factors determine the amount of drip on thawing. Of these, the most important is the ultimate pH of the meat. It was first shown 60 years ago (Empey, 1933) that a high ultimate pH in meat was associated with a lower than normal percentage of drip after freezing and thawing, and this was confirmed with carcass meat (Howard and Lawrie, 1956). That differences in the proteins of different muscles are also a factor is clear from the fact that, at a given ultimate pH, bovine *l. dorsi* exudes about twice as much drip as the *psoas* (Lawrie, 1959). Again, on thawing, frozen conditioned meat tends to drip less than meat which has not been aged, possibly because the breakdown processes involved increase the intracellular osmotic pressure and thus enhance water-holding capacity.

When freezing is applied to muscles in which *post mortem* glycolysis is still proceeding, and before the onset of *rigor mortis*, the system may be frozen whilst the concentration of ATP is still relatively high. On thawing, such muscles will contract extremely (for example, to 30% of prefreezing length), exude massive amounts of drip and produce exceedingly tough

Table 1.4 Approximate times (months) for appearance of distinct rancidity in fat or discoloration of lean in unwrapped meat at various storage temperatures (after Dahl, 1958)

Meat	−8°C	−15°C	−22°C	−30°C
Beef (rib roasts)	3	6	12	–
Pork (loin without rind and back fat)	–	3	6	12

meat. This is the phenomenon of 'thaw rigor' (Chambers and Hale, 1932). It is caused by an abnormally powerful stimulation of the contractile myosin ATPase, this being due, in turn, to a combination of high concentrations of calcium ions (released into the system through the disruption caused by ice formation) and of ATP (which has been protected from normal attack during *post-mortem* glycolysis by immobilisation of the system on freezing) (Bendall, 1973). 'Thaw rigor' can be avoided if the frozen pre-rigor meat is held at a few degrees below the freezing point before final thawing when, although there is sufficient thawing to cause the accumulated ATP to break down, the muscles cannot contract because they are held rigid by the remaining ice (Marsh and Thompson, 1958).

Even in the frozen state, meat will deteriorate through non-enzymic oxidation of lipids. The extent depends on species, pork being more susceptible than beef, and on the temperature of frozen storage (Table 1.4). Packaging film, if sufficiently impermeable to moisture, will limit desiccation, oxidation and discoloration; it must be sufficiently robust to resist loss of functionality in a freezing environment. Materials are now available which soften on heating and may thus be moulded to the meat surface, leaving no spaces in which local oxidation can occur. The films are transparent and, since they have considerable permeability to oxygen, it is possible to retain the bright red colour of oxymyoglobin on the meat surface for as long as a year at −20°C if the product is kept in the dark (Taylor, 1985).

It must be emphasised that freezing does not confer absolute safety from microbial poisoning. Spores may survive and germinate in the thawed meat if there is a delay in consumption. Moreover, if toxins have been produced by the growth of pathogens before freezing, these could affect consumers of the product subsequently.

Temperatures above the optima for microbial growth can also be employed to preserve meat. This principle is the basis of canning and in cabinets used to dispense hot food. In the former it is important to ensure that, once the required heat has been administered, the meat is thereafter kept sterile by effective sealing of the container. Because it is essential that the deepest location receives the minimum heat to inactivate any bacteria present, exterior locations may be overprocessed whereby the meat proteins will suffer some damage, and the heated product will tend to resemble the cooked meat rather than the fresh commodity. Thermal processing at a level to kill all non-sporing bacteria and all spores capable of germinating if the product is stored without refrigeration, is the purpose of commercial sterility (Ayers *et al.*, 1980). Because the proteins of some meats such as pork are relatively labile on heating, a less intense heating process, pasteurising, is applied to them. Pasteurised or semipreserved meat will not remain unchanged unless kept at chill temperatures or unless they contain curing salts of which the ingredients will lower microbial resistence to heat. Insufficient heat processing can lead to can swelling (due to the survival of micrococci, bacilli, coliforms or clostridia) or, more rarely, to putrefaction (due to *Achromobacter* or *Pseudomonas*) and growth of pathogens (due to *Salmonella*) (Mossel *et al.*, 1975; Hobbs and Gilbert, 1978). Whether thermal processing involves commercial sterility or pasteurising, it is essential that *Cl. botulinum* is destroyed or prevented from developing. If thermal processing is carried out in accordance with prescribed schedules, canned meats may be kept for very long periods without deterioration in eating quality — albeit that, over a century, the lysine present may become less biologically available.

Relatively recently, flexible bags, stable to sterilising doses of heat and capable of being hermetically sealed, have made heat transfer much more effective than is possible with rigid cans. The severity of the heat required to achieve sterility is thereby less and damage to the quality of the meat minimal. The shelf-life is comparable to that of a frozen product, and in many cases the product can be heated for consumption within the bag.

1.5.2 Moisture control

The removal of available moisture is another major strategy for inhibiting microbial growth in meat. Such can involve lowering the water content, as in dehydration and freeze-drying, or making the water present unavailable to microorganisms, as in curing and intermediate moisture technology. Empirical drying methods were employed from the earliest times and traditional dried products, such as charqui and biltong, are still made although, to most palates, they lack much of the organoleptic appeal of fresh meat. During the Second World War, a method for producing acceptable dehydrated meat on a commercial scale was devised. To achieve rapid heat transfer it was necessary to mince the meat, but this led to uneven drying, with case hardening of the exterior, and moisture retention in the interior, of the pieces. By first cooking the meat in slices, however, then dehydrating the cooked mince, a uniform dehydrated product could be prepared. Since dehydrated meat would be consumed after cooking, the precooking operation was not a disadvantage, although it was necessary to add back the juices exuded from the meat on cooking to the cooked mince, before dehydration. Provided the temperature of the

Table 1.5 The effect of plate temperature on various characteristics of freeze-dried meat after rehydration (reproduced from Lawrie (1991) by kind permission of Pergamon Press Ltd).

| Characteristic | Control (frozen) | Temperature of plate (°C) | | | | |
		20	40	60	80	100
Total moisture (%)	77.2	75.4	75.9	75.5	76.8	75.3
'Bound' water (%)	65.3	63.8	63.9	63.1	62.8	53.8
Meat solids (%) in reconstitution water	12.3	10.1	9.9	10.9	10.3	5.4

hot air used to remove moisture was kept below 70°C, the meat after rehydration was not markedly different from comparable cooked fresh meat (Sharp, 1953).

Dehydrated meat has a mean residual moisture content of 7.5%, which is sufficiently low to prevent bacterial growth; if it is above 10%, moulds, such as *Penicillium* and *Aspergillus* may develop. In the absence of oxygen, the main deteriorative change of dehydrated meat on storage is due to the Maillard reaction — browning and bitterness resulting. In the presence of oxygen, pigment fading and lipid oxidation and rancidity occur. If compressed to exclude air pockets, and sealed in cans under nitrogen, such dehydrated meat could be kept for many years at moderate storage temperatures without loss of quality (Sharp, 1953).

The need to mince and cook the meat in order to achieve effective heat transfer in the dehydration process were serious disadvantages. Moreover, it had been appreciated for some time that heat-sensitive pharmaceutical substances could be dehydrated without loss of activity if they were first frozen and the ice removed by sublimation under vacuum. A procedure for the production of freeze-dried meat on a large scale was developed in the 1950s. Since it incorporated heated plates to assist heat exchange during the change of state, it became known as the accelerated freeze-drying process (AFD) (Rolfe, 1958). It had advantages over dehydration *per se* since it could dry whole steaks of raw meat to a residual moisture content of less than 2% in a few hours. When operated under optimum conditions, i.e. with the temperature of the heated plates between 20 and 30°C (Table 1.5) the freeze-dried meat rehydrated easily and almost completely and was virtually indistinguishable from corresponding fresh meat. Its colour was bright red and the proteins little altered. Damage to proteins is minimal with freeze-drying because the cell water is frozen *in situ* and removed without distortion of the structure by sublimation, whereas in air-drying water is removed by slow translocation during which cell walls are damaged and proteins denatured by local concentrations of muscle salts. The tenderness of freeze-dried meat, although very satisfactory, can be enhanced even further by the use of meat of high ultimate pH (Table 1.6; after Penny *et al.*, 1963). As in the case of dehydrated meat, the deterioration of freeze-dried meat on storage is mainly due to Maillard-

Table 1.6 The sum of the tenderness rankings* given for control (low ultimate pH) and adrenaline-treated (high ultimate pH) beef before and after freeze dehydration (after Penny *et al.*, 1963)

Muscle	Control		Adrenaline treated	
	Frozen	Freeze-dried	Frozen	Freeze-dried
Semimembranosus	23	24	12	16
Biceps femoris	26	25	11	15
Psoas major	23	25	12	14
L. dorsi (lumbar)	19	21	19	17
L. dorsi (thoracic)	26	25	11	12
Pectoralis profundus	25	27	10	16
Total	142	147	75	90

* Low numbers represent desirable degrees of tenderness and *vice versa*: eight tasters were involved.

type browning or fat rancidity; these changes can be largely prevented by pre-packaging.

In the USA, freeze-dried steaks are available in moisture-proof packs in which are enclosed dried preparations of proteolytic enzymes. These can be added to the rehydration water to ensure tenderness. Economic factors have prevented the full development of freeze-drying as a preservative process for meat, especially as the commodity can be excellently preserved by modern refrigeration techniques.

The observation that the addition of salt or sugar to meats not only prolonged the acceptable storage life but also developed attractive organo-leptic features, was evidently appreciated in ancient times, as shown by the illustrated friezes of Egypt and Babylon. Adventitious impurities in the salt (usually sodium chloride), in particular nitrates and nitrites, were substan-tially responsible for the organoleptic properties and resistance to microbial spoilage of such cured products.

Although salted beef and mutton used to be available, they were somewhat unappetising and, when alternative procedures developed, they lost popularity. On the other hand, the curing of pork yielded a product which had attractive features differing markedly from those of the fresh commodity. As a result, cured pork products have survived in parallel with other preserved forms of pork. In curing, dressed sides of pork used to be sprinkled with salt. Later they were immersed in tanks of brine; later still, the curing process was made more effective by adding brine by arterial pumping or multiple needle injection before sides were submerged. Because the top of the hind limb was thick and difficult to cure, it was frequently removed and cured separately by a variety of salt or sugar mixtures and yielding hams of subtly different character. (When cured on the side, the top of the hind limb is known as gammon.)

Much bacon was produced by the Wiltshire cure or its variants (Callow, 1934). After butchering into sides, the backbone was removed and the sides cooled and trimmed. They were then artery pumped or injected with brine under pressure (pump pickle) which contained 25–30% sodium chloride and 2.5–4.0% potassium or sodium nitrate. The injected sides were then placed 12 deep in tanks and battened down tightly. The sides were then submerged in brine (tank pickle) containing 20–28% sodium chloride and 3–4% potassium nitrate. After 4–5 days, the sides were removed from the tank and held for a further 7–14 days in chilled cellars to mature, when the sodium chloride, nitrate and the nitrite into which the latter was converted, became more evenly distributed in the bacon. It was essential in the original curing procedure to ensure that the tank pickle contained a viable population of halophilic bacteria since it was due to their action that the nitrate was converted into nitrite, the latter being largely responsible for the development of the typical colour and flavour of bacon and, more importantly, for specific inhibition of *Cl. botulinum*, for which a minimum residual concentration of 100 ppm is necessary (Anon, 1974). If bacon is sterilised by 20–30 kGy of ionising radiation, however, safe storage can be assured without nitrite, but 20–40 ppm is still needed for colour and flavour development (Wierbicki and Heligman, 1980).

Advances in curing have permitted a considerable acceleration of the process. If the pieces of pork are massaged or 'tumbled' in rotating drums in contact with 0.6% of their weight of salt, salt distribution is rapid and curing can be accomplished in 24 hours. It is also possible to produce bacon in a sterile cure. In this, thin slices of pork are passed for 2–15 min through a bath containing 8–10% sodium chloride and 0.02% sodium nitrite, no microorganisms being thus required to produce the latter from nitrate. This procedure also yields a uniform product in 24 hours. The typical colour and flavour of bacon develop during distribution (Holmes, 1960). Elevated temperatures will also increase the velocity of salt penetration but this, of course, involves the risk of microbial spoilage.

The presence of various contaminating microorganisms in the curing solutions can cause souring (*Lactobacillus*), putrefaction (*Vibrio*) and excessive sweetness (*Bacillus*) in the bacon (Leistner, 1960). If the pork has an elevated ultimate pH, the conversion of nitrate to nitrite by the halophils during a few days storage at chill temperature (Jolley, 1979) can lead to unacceptably high concentrations. Current policy is to limit the residual nitrite to below 200 ppm to avoid the possibility of carcinogenic nitrosoamines being formed by interaction of the nitrite with secondary and tertiary amines in the pork.

Once the bacon has been produced, its storage life can be increased (and its flavour altered) by the bacteriostatic action of smoke produced from the slow combustion of hardwood sawdust. The effective components of the smoke include fatty acids, formaldehyde and various phenolic substances. Unfortunately carcinogenic benzpyrenes are also present.

On storage, cured meats deteriorate in the first instance, through discoloration; secondly, the lipids undergo oxidative rancidity; and, thirdly, microbial changes occur. The last has become more significant with pre-packaged bacon. In the presence of oxygen, light accelerates discoloration through the production of brown metmyoglobin. The salt content of cured meats causes the lipids to oxidise more rapidly than those in the fresh commodity, although smoking has antioxidant activity. Ascorbic acid, sometimes added to the curing ingredients to protect the bacon colour, is also a pro-oxidant of fat (Watts, 1954). Although vacuum packaging, by excluding oxygen, prevents discoloration and limits fat oxidation, the extra handling during its preparation makes the product liable to be more contaminated than when slices were cut from bacon sides and joints at time of sale. The atmosphere within the vacuum packs may be altered by such contaminants and by indigenous microflora. Traces of residual oxygen could be converted to carbon dioxide which would inhibit the normal halophiles and cause their replacement by organisms which could change the appearance, flavour and safety of the products. In bacon, spoilage bacterial numbers reach a maximum, however, some days before eating quality deteriorates (Kitchell and Ingram, 1963). Above 20°C, *Staph. aureus* will tend to dominate the microflora and putrefactive odours will arise if the temperature is *ca.* 30°C, as might occur within packs exposed to sunlight (Cavett, 1962).

The high salt content in cured products, on the one hand, and the difficulty of rehydrating dehydrated products, on the other, has fostered the development of so-called intermediate moisture meats (Brockmann, 1970). In these, the water is lowered to the point at which bacteria will not grow, even at high ambient temperature, without lowering the a_w to the point of unpalatability. In this process, lean meat is cut into 1 cm cubes which are immersed in 1.5 times their weight of infusing solution. The latter consists of 10% sodium chloride, 0.5% of an antimycotic and sufficient glycerol (33–40%) to lower the a_w to 0.82–0.60. The mixture is then heated at 70°C in cans. The surfaces of the meat pieces are dried and these are stored in impermeable plastic baths. Intermediate moisture meat can be kept at 38°C and still remain acceptable for several months, although textural changes gradually occur — firstly a tenderising due to slow breakdown of collagen, later hardening as Maillard-type compounds form (Ledward, 1981). Intermediate moisture meat (and other intermediate moisture products) have applications in space travel and in other situations when stability, palatability and convenience are essential.

1.5.3 Direct microbial inhibition

As an alternative to controlling the growth of microorganisms by exposing them to suboptimal temperatures or moisture levels, they can be inhibited

Table 1.7 Doses of ionizing radiation required for various microbiological effects (after Brynjolfsson, 1980)

Effect	Dose (kGy)
10^6 reduction in number of fungi	1–10
10^6 reduction in number of vegetative bacteria	0.5–19
10^6 reduction in number of *Salmonella* spp.	2–7
10^6 reduction in number of spores and dried bacteria	8–25
10^6 reduction in number of viruses	10–40
Commercial sterility (10^{12} reduction in number of *Cl. botulinum*)	45
Inhibition of enzymes	1000

or killed by more direct action, using ionising radiation, antibiotics or by various chemicals of general toxicity to biological tissues.

Ionising radiation signifies the production of ions from the molecules of the material against which the rays are directed. The ions then react to form free radicals and cause other chemical changes. The ionising rays may be particulate (e.g. electrons, neutrons, α-particles) or electromagnetic (e.g. X-rays or γ-rays). Of the many varieties, however, only high-energy electrons or soft X-rays from generators, or γ-rays from ^{60}Co, are of practical use in treating foods (Hannan, 1955). It was first appreciated 60 years ago that their efficacy in killing bacteria was due to the damage the ions caused to the deoxyribose nucleic acid of the reproductive centres. These very large molecules are particularly susceptible and indeed there is an inverse relationship between the size of molecules and the dose of ionising radiation needed to inactivate or destroy them. This means that a dose sufficient to kill bacteria will be much less than that required to damage even the relatively large molecules of food proteins. The dose of ionising radiation was originally measured in rads; it is now measured in grays (Gy). A gray is the absorption of 10^5 ergs per gram of the target.

Even among microorganisms there is a range of susceptibility (Table 1.7). It will be seen that to destroy all microorganisms in foods, including viruses, requires 60 kGy. Commercial sterility (i.e. achieving a 10^{12} reduction in the numbers of *Cl. botulinum*) requires not less than 45 kGy. Such a dose changes about 0.2% of the proteins in the food (Brynjolfsson, 1980), but it cannot be employed to sterilise meat since extremely persistent volatiles are produced, in minute quantities, giving an unacceptable 'wet dog' odour. Thus, despite the possibility of sterilising meat within packs in which they can be stored more or less indefinitely without microbial spoilage and without the need for refrigeration, ionising radiation is likely to be employed for significantly extending the storage life of meat at chill temperatures, using low doses which have negligible effect on the molecules of the proteins present. Such pasteurising doses (*ca.* 5 kGy) will destroy food poisoning microorganisms (e.g. *Salmonella, Staphyloccocus, Campylobacter jejuni*) and psychrophilic spoilage organisms, within which there are differences of radiation susceptibility. Thus a dose of 2–3 kGy will cause the replacement of *Pseudomonas* and *Achromobacter* by yeasts (Thornley *et al.*, 1960).

An important factor determining the radiation resistance of microorganisms is the presence or absence of oxygen. Thus, the removal of oxygen will increase the resistance of *E. coli* threefold (Niven, 1958). Of the psychrotrophs, *B. thermosphacta* is relatively resistant to pasteurising doses and, even if only small numbers survive, they may cause a sour odour. Meat treated with 0.5–10 kGy can be held at 0–5°C for ten times longer than non-irradiated meat before microbial spoilage ensues (Niven, 1963). Commercial-scale studies on meat transported to UK from Australia and New Zealand (Rhodes and Shepherd, 1966) showed that 4 kGy would delay microbial spoilage on the surface of lamb carcasses (held in air-free packs), for more than 8 weeks at 1°C. (It should be appreciated, that whereas pasteurising by heat will destroy the toxin of *Cl. botulinum*, a pasteurising dose of ionising radiation will not.).

Because of its comminuted nature and the mode of its preparation, ground beef is highly susceptible to microbial spoilage. Since it constitutes *ca.* 30% of Canadian expenditure on beef, its perishability is clearly an important problem. In a recent investigation, however, it was demonstrated that doses of 1, 2.5 or 5 kGy, delivered by γ-rays from ^{60}Co, could extend the useful storage life of the product at 4°C by 4, 10, or 15 days beyond that of non-irradiated controls (Lefebvre *et al.*, 1992).

Although pasteurising doses of ionising radiation cause so little damage to food molecules, 0.25–1 kGy may initiate oxidation of lipids and, because of the prolonged period of storage possible without microbial spoilage, this may lead to the accumulation of tallowy flavours (Lea *et al.*, 1960), an effect which can be largely prevented, however, by using an oxygen-impermeable packaging film.

Apart from its usefulness in preserving chilled meat for human consumers, ionising radiation at doses of 6.5 kGy will eliminate *Salmonella* from imported meat used in preparing petfood — and thus protect consumers indirectly. Fears that, despite its negligible effect on food constituents, ionising radiation might produce highly potent carcinogens (albeit in minute quantities) have been dispelled by the absence of any untoward effects in long-term animal experiments (Elias, 1985). Indeed in 1981 an expert committee of FAO/IAEA/WHO pronounced that there was no hazard to consumers from foods receiving doses of up to 10 kGy, provided the incident energy of the rays did not exceed 5 MeV (if electromagnetic rays were used) or 10 MeV (if electrons were used).

The observation by Fleming in 1929 that some microorganisms produced substances inimical to the growth of others — antibiotics — and their very low toxicity to human beings soon led to their use in medicine to combat bacterial infections (Abraham *et al.*, 1941). The feasibility of using them to preserve foods was appreciated somewhat later (Tarr *et al.*, 1952). Since antibiotics, at practical concentrations, do not sterilise but only delay bacterial growth, they are most effective when bacterial numbers are relatively low. Bacteria differ in their susceptibility and, as a result, antibiotics may cause the development of resistant microorganisms in meat after susceptible bacteria have been eliminated and thus no longer compete with them. Broad-spectrum antibiotics, such as the tetracyclines, which are effective against both Gram-negative and Gram-positive bacteria (but not yeasts or moulds) were thus favoured. The initial potential of antibiotics has not been fulfilled, however, since it eventually became apparent that populations of microorganisms included resistant strains which could survive and subsequently multiply given appropriate conditions. It was thus deemed wise to limit the use of antibiotics, e.g. nisin, in foods to those which are not employed in combatting human disease, and those only in meats when these have been heated sufficiently to destroy *Cl. botulinum*. Insofar as antibiotics suppress those spoilage organisms which produce detectable evidence of deterioration to the consumer, and thus remove competition with pathogens of which there could be no overt signs, antibiotics have tended to be used to extend the storage life of *chilled* meat products. Thus the storage life of ground beef can be extended to over a week even at 10°C by the incorporation of 0.5–2 ppm of chloramphenicol, chlortetracycline or oxytetracycline (Goldberg *et al.*, 1953) and that of pork sausages trebled. If the temperature of such meat products rises to 15°C, however, antibiotic-resistant pathogens may be able to multiply (e.g. *S. typhimurium*; Hobbs, 1960).

An additional potential disadvantage in using antibiotics is toxicity or allergy to consumers from residues which have survived cooking, although generally the level of antibiotics originally present in meat largely diminishes during the period until consumption (Weiser *et al.*, 1954).

Antibiotics are up to 100-fold more effective than corresponding concentrations of the chemicals which have been used to preserve meats. Many of these are as toxic to human consumers as they are to microorganisms, and their abuse in the 19th century was largely responsible for the establishment of the comprehensive Food Laws which now protect the customer — usually most effectively. Under the UK Preservatives in Food Regulations of 1962 only seven such chemicals are now permitted as preservatives, and only one of these, sulphur dioxide, is permitted in meat (up to 450 ppm in sausage meat).

Various naturally-occurring essential oils have preservative components, e.g. eugenol in cloves and allylisothiocyanate in mustard seed. Thus Shelef *et al.* (1980) demonstrated that 0.5% of sage or rosemary was bactericidal. Such materials have been employed to extend the storage of meat products (McNeill *et al.*, 1973). A number of chemicals with preservative properties are specifically excluded from the UK Preservatives in Food Regulations.

These include ingredients traditionally used in curing, such as sodium chloride and sugar, lactic acid (which is produced in meat during *post mortem* glycolysis) and acetic acid.

Various organic acids have long been used in preparing marinaded meats, i.e. meats pickled in wine or vinegar, with herbs and spices. Since there has been an increase in the range of marinaded meats available, Gault (1991) reviewed their characteristics in general, and reported a detailed investigation on the nature of those involving acetic acid. In his study he overcame the variability caused by the slow inward diffusion of the acid by using small discs (30-mm diameter, 10-mm thickness, 8-g mass) prepared from a number of post-rigor bovine muscles. These were immersed in 50 ml of a series of acetic acid solutions, of which those of between 2.5 and 5 mM proved most effective. The marinades were maintained at 4°C for 48 hours with constant agitation. The pH of the meats so treated ranged from 5.0 down to 3.0. Below pH 4.5 the meat swelled increasingly to a maximum at pH 3.4, the tenderness of the cooked meat increasing concomitantly (Gault, 1985).

Of the various organic acids which have been used in marinading meat, acetic acid is the most effective against microorganisms. Thus, *Lactobacillus brevis* is inhibited from growth at pH 4.0 when the acidulant is acetic acid, whereas with lactic acid, growth occurs down to pH 3.7. It is important to ensure that the initial microbial load is low since organic acids have a decreasing effect in heavily contaminated meats.

1.5.4   Fermentation

Thousands of years before microorganisms (or antibiotics) were known, the capacity of some to thrive in foods by producing metabolites injurious to competing spoilage species was inadvertently exploited by human beings in preserving commodities (Brothwell and Brothwell, 1969). Such preservation depends on the anaerobic breakdown of carbohydrates to yield various alcohols and organic acids (formic, acetic, propionic or lactic), i.e. fermentation by non-pathogenic microorganisms. These chemicals, their nature depending on the food, the organism and the environment, inhibit the growth of microbial species which make foods aesthetically unacceptable or toxic. The conditions which fostered preservation (i.e. the growth of the effective fermentative microorganisms) were originally established by empirical observation.

In respect of meat, fermentation has been employed to preserve or enhance the organoleptic attributes of comminuted products such as sausages, in which the *milieu* already differs from that of fresh meat due to the presence of salts or sugars. The latter circumstance selects the type of fermentative microorganisms which will grow at relatively high osmotic pressure. Whereas indigenous microorganisms were responsible for the fermentation traditionally, and are still encouraged in some products, desired organoleptic attributes can now be assured by adding cultures of predetermined composition, under strictly controlled operating conditions. In a detailed study of the structure of fermented meats, using scanning electron microscopy, Katsaras and Budras (1992) have demonstrated the interdependence of the configuration of the protein matrix, the salt concentration and the growth of the fermenting microorganisms, and thus the importance of the manner and extent of meat comminution.

The antimicrobial action of lactic acid, which is naturally produced in muscle during its *post mortem* conversion to meat, has been mentioned above. Lactic acid is also the organic acid which is most frequently produced by the microbial fermentation of meats. Fermented sausages have a long storage life (*ca.* 1–2 years), due to the added salt, to the lactic acid produced *in situ* by lactobacilli (or other starter culture) during the early stages of the storage period, to nitrite (formed from nitrate by halophilic micrococci, as in bacon curing), and to the drying which occurs as storage progresses. In view of the relatively recent appreciation that *Yersinia enterocolitica* and *Listeria monocytogenes* can endanger human health, it is reassuring to note that the organisms which flourish in fermented meat products can inhibit their growth (Raccach and Henningsen, 1984; Schillinger *et al.*, 1991).

Whereas in traditional meat fermentation the changes desired developed over a considerable time, faster fermentation is being currently sought (Bacus, 1984). Clearly there is a concomitant necessity to ensure that the accelerated processes are still precisely controlled, especially since a greater range of flavour attributes is now demanded to satisfy more discriminating palates. Thus the Australian Code of Manufacturing Practice of 1982 for dry and semi-dry sausages, requires the pH to fall to 5.2 within 48 hours and the fermentation temperature to be kept below 25°C. In the same year a task force, sponsored by the American Meat Institute and other bodies, recommended that, in good manufacturing practice, the addition of lactic acid-forming bacteria should reduce the pH of dry and semi-dry sausage mix to ≤ 5.3 within times determined by the fermentation temperature (e.g. 80 hours at 24°C, 18 hours at 43°C; Bacus, 1984). These criteria are presumed to prevent the growth of *Staph. aureus*, at temperatures which would otherwise favour this organism.

In a comparison of starter cultures relevant to such criteria, Coventry and Hickey (1991) demonstrated that the growth of neither *Lactobacillus plantarum* nor *Pediococcus pentosaceus*, which are otherwise effective, prevented the development of high levels of competing non-starter microorganisms in salami, although the more psychrotrophic *P. pentosaceus* ensured a greater dominance of the starter organisms and thus minimized spoilage. As Smith and Palumbo (1981) have pointed out, the techniques of meat fermentation are now being applied to various non-fermented products to accelerate flavour development, extend shelf-life and control the growth of pathogens. Starter cultures have been increasingly used to reduce nitrite concentration in cured meats and to lessen spoilage by metabolically active microbial contaminants (Bacus, 1984).

There is widening interest in fermentation as a mode of meat preservation and as a means of increasing the diversity of its products, as subsequent chapters in this volume attest.

References

Abraham, E.P., Chain, E.B., Fletcher, C.M., Florey, H.W., Gardner, A.D., Heatley, N.G. and Jennings, M.A. (1941) Further observations on penicillin. *Lancet*, **ii**, 177–188.

Analytical Methods Committee (1991) Nitrogen factors for pork: a reassessment, *Analyst*, **116**, 761–766.

Anon (1974) *Ann. Rept. Meat Res. Inst.*, Bristol, 1972–73, pp. 11–12.

Anon, M.C. and Calvelho, A. (1980) Freezing rate effects on the drip loss of frozen beef. *Meat Sci.*, **4**, 1–14.

Ayers, J.C., Mundt, J.O. and Sandine, W.E. (1980) *Microbiology of Foods*, pp. 94–102, 417–431. W.H. Freeman, San Francisco.

Bacus, J. (1984) Update: meat fermentation, 1984. *Food Technol.* **38**(6), 59–63.

Bailey, A.J. (1974) Tissue and species specificity in the cross-linking of collagen. *Path. Biol.*, **22**, 675–680.

Bailey, A.J. and Light, N. (1989) *Connective Tissue in Meat and Meat Products*, pp. 6–49. Elsevier Applied Science London.

Bailey, A.J., Robins, S.P. and Balean, G. (1974) Biological significance of the intermolecular cross-links of collagen. *Nature London*, **251**, 105–109.

Bailey, C. and Cox, R.P. (1976) *The Chilling of Beef Carcass*. Institute of Refrigeration, London.

Barlow, J. and Kitchell, A.G. (1966) A note on the spoilage of prepacked lamb chops by *Microbacterium thermosphactum J. Appl. Bact.*, **29**, 185–188.

Barnes, E.M. and Ingram, M. (1956) The effect of redox potential on the growth of *Cl. welchii* strains isolated from horse muscle. *J. Appl. Bact.* **19**, 117–128.

Bate-Smith, E.C. and Bendall, J.R. (1949) Factors determining the time course of *rigor mortis. J. Physiol.*, **110**, 67–74.

Bendall, J.R. (1973) *Post mortem* changes in muscle. In *The Structure and Function of Muscle*, 2nd edn. Ed. G.H. Bourne, pp. 244–309. Academic Press, New York.

Bendall, J.R. (1980) The electrical stimulation of carcasses of meat animals. In *Developments in Meat Science*. 1. Ed. R.A. Lawrie, pp. 37–59. Elsevier Applied Science, London.

Bernard, C. (1877) *Leçons sur la Diabète et la Glycogenèse Animale*, p. 426. Baillière, Paris.

Briskey, E.J. (1964) Etiological status and associated studies of pale, soft, exudative porcine musculature. *Adv. Food Res.*, **13**, 90–178.

Brockmann, M.C. (1970) Development of intermediate moisture foods for military use. *Food Technol., Champaign*, **24**, 896–900.

Brooks, J. (1935) The oxidation of haemoglobin to metmyoglobin by oxygen. II. The relation between the rate of oxidation and the partial pressure of oxygen. *Proc. Roy. Soc. B*, **118**, 560–577.

Brothwell, D. and Brothwell, P. (1969) *Food in Antiquity*, p. 195. Thames & Hudson, London.

Brown, M.H. (1982) *Meat Microbiology*. Elsevier Applied Science, London.

Brynjolfsson, A. (1980) Food irradiation in the United States. *Proc. 26th Meeting Eur. Meat Res. Workers, Colorado Springs*, **1**, 172–177.

Callow, E.H. (1934) *Bacon curing: the Dry-salt and Tank Curing of Wiltshire Sides*. Food Investigation Board, London, Leaflet No. 5.

Callow, E.H. (1937) The ultimate pH of muscle tissue. *Annu. Rept. Food Invest. Bd.*, London pp. 49–51.

Carse, W.A. (1973) Meat quality and the acceleration of *post mortem* glycolysis by electrical stimulation. *J. Food Technol.*, **8**, 163–166.

Castéllani, A.G. and Niven, C.F. Jr. (1955) Factors affecting the bacteriostatic activity of sodium nitrite. *Appl. Microbiol.*, **3**, 154–159.

Cavett, J.J. (1962) The microbiology of vacuum-packed, sliced bacon. *J. Appl. Bact.*, **25**, 282–289.

Chambers, R. and Hale, H.P. (1932) The formation of ice in protoplasm. *Proc. Roy. Soc. B.*, 110, 336–352.

Citoler, P., Benitez, L. and Maurer, W. (1966) Autoradiographische Untersuchung der Protein-syntheserate in roten und weissen muskelfasern. *Exp. Cell Res.*, **45**, 195–205.

Cohen, B. and Clark, W.M. (1919) The growth of certain bacteria in media of different hydrogen ion concentration. *J. Bact.*, **4**, 409–426.

Cook, G.A., Love, E.F.G., Vickery, J.R. and Young, W.G. (1926) Studies on the refrigeration of meat. I. Investigations into the refrigeration of beef. *Aust. J. Exp. Biol. Med. Sci.*, 3, 15–31.

Cosnett, K.S., Hogan, D.J., Law, N.H. and Marsh, B.B. (1956) Bone taint in beef. *J. Sci. Food Agric.*, 7, 546–551.

Coventry, J. and Hickey, M.W. (1991) Growth characteristics of meat starter cultures. *Meat Sci.*, **30**, 41–48.

Dahl, O. (1958) Freezer storage of beef at different temperatures. *J. Refrig.*, 1, 170–171.

DeDuve, C. and Beaufay, H. (1959) Tissue fractionation studies. 10. Influence of ischaemia on the state of some bound enzymes in rat liver. *Biochem. J.*, 73, 610–616.

Dolman, C. (1957) The epidemiology of meat-borne diseases. In *Meat Hygiene*, WHO Monograph No. 33, pp. 11–108, FAO, Rome.

Dransfield, E. (1992) Modelling *post mortem* tenderisation. III. Role of calpain I in conditioning. *Meat Sci.*, **31**, 85–94.

Dransfield, E., Jones, R.C.D. and MacFie, H.J. (1981) Tenderisation in *m. longissimus dorsi* of beef, veal, rabbit, lamb and pork. *Meat Sci.*, **5**, 139–147.

Dutson, T.R., Smith, G.C. and Carpenter, Z.L. (1980) Lysosomal enzyme distribution in electrically stimulated ovine muscle. *J. Food Sci.*, **45**, 1097–1098.

Eddy, B.P., Gatherum, D.P. and Kitchell, A.G. (1960) Bacterial metabolism of nitrate and nitrite in maturing bacon. *J. Sci. Food Agric.*, 11, 727–735.

Elias, P.S. (1985) Irradiation preservation of meat and meat products. In *Developments in Meat Science. 3*. Ed. R.A. Lawrie, pp. 115–153. Elsevier Applied Science, London.

Empey, W.A. (1933) Studies on the refrigeration of meat. Conditions determining the amount of drip from frozen and thawed muscle. *J. Soc. Chem. Ind.*, **52**, 230T–236T.

Empey, W.A. and Scott, W.J. (1939) Investigations on chilled beef. I. Microbial contamination acquired in the meatworks. *CSIRO Bull.* No. 126.

Galbraith, N.S., Barrett, N. and Sockett, P.N. (1987) The surveillance of food borne infection: the role of the communicable disease surveillance centre. *BNF Bull.*, **12**, 21–31.

Gault, N.F.S. (1985) The relationship between water-holding capacity and cooked meat tenderness in some beef muscles as influenced by acidic conditions below the ultimate pH. *Meat Sci.*, **15**, 15–30.

Gault, N.F.S. (1991) Marinaded meat. In *Developments in Meat Science. 5*. Ed. R.A. Lawrie, pp. 191–246. Elsevier Applied Science, London.

Gauthier, G.F. (1969) On the relationship of the ultrastructure and cytochemical factors to color in mammalian skeletal muscle. *Zellforschung*, **95**, 462–482.

George, J.C. and Naik, R.M. (1958) Relative distribution of the mitochondria in the two types of fibres in the *pectoralis* muscle of the pigeon. *Nature (London)*, **181**, 782–783.

Gergely, J., Pragey, D., Scholtz, A.F., Seidel, J.C., Sréter, F.A. and Thompson, M.M. (1965) Comparative studies on white and red muscle. In *Molecular Biology of Muscular Contraction*. Eds S. Ebashi, F. Oosawa, T. Sekine and Y. Tonomura, pp. 145–149. Ikagu Shoin, Tokyo.

Gigiel, A.J. and James, S.J. (1984) Electrical stimulation and ultrarapid chilling of pork. *Meat Sci.*, **11**, 1–12.

Gill, C.O. (1989) Packaged meat for prolonged chill storage: the CAPTECH process. *Br. Food J.*, **91**(7), 11–15.

Gill, C.O. and Penney, N. (1986) Packaging conditions for extended storage life of chilled, dark firm dry beef. *Meat Sci.*, **18**, 41–54.

Goldberg, H.J., Weiser, H.H. and Deatherage, F.E. (1953) Studies on meat. IV. The use of antibiotics in the production of fresh beef. *Food Technol.*, **7**, 165–166.

Goldspink, G. (1970) Morphological adaptation to growth and activity. in *Physiology and Chemistry of Muscle as a Food*. Vol. II. Eds E.J. Briskey, R.G. Cassens and B.B. Marsh, pp. 521–536. University of Wisconsin Press, Madison, WI.

Goll, D.E., Stromer, M.H., Robinson, R.M., Temple, J., Eason, B.A. and Busch, W.H. (1971) Tryptic digestion of muscle components simulates many of the changes caused by *post mortem* storage. *J. Anim. Sci.*, **33**, 963–982.

Grau, F. (1988) *Campylobacter jejuni* and *Campylobacter hyointestinalis* in the intestinal tract and on the carcasses of calves and cattle. *J. Food Protect.*, **51**, 857–861.

Greer, G.G. and Murray, A.C. (1988) Effects of pork muscle quality on bacterial growth and retail case life. *Meat Sci.*, **24**, 61–72.

Gunther, H. (1921) cited by D.M. Needham (1971) in *Machina Carnis*, p. 451. Cambridge University Press, Cambridge.

Haines, R.B. (1937) *Microbiology in the Preservation of Animal Tissues*. Spec. Rept. Food Invest. Bd., London, No. 61. HMSO, London.

Hannan, R.S. (1955) *Ionizing Radiation*. Spec. Rept. Food Invest. Bd., London, No. 61. HMSO, London.

Hansom, J. and Huxley, H.E. (1953) Structural basis of the cross-striations in muscle. *Nature (London)*, **172**, 530.

Hobbs, B.C. (1960) The antibiotic treatment of poultry in relation to *S. typhimurium*. *Proc. 6th Meeting Europ. Meat Res. Workers, Utrecht*, Paper 32.

Hobbs, B.C. and Gilbert, R.J. (1978). *Food Poisoning and Food Hygiene*, 4th edn., pp. 54, 94. Edward Arnold, London.

Holmes, A.W. (1960) Unilever, UK. Brit. Pat. No. 848014.

Howard, A. and Lawrie, R.A. (1956) *Studies on Beef Quality, I–III*. Spec. Rept. Food Invest. Bd., London, No. 63, pp. 18–51. HMSO: London.

Huxley, H.E. (1963) Electron microscope studies on the structure of natural and synthetic protein filaments from striated muscles. *J. Mol. Biol.*, **7**, 281–308.

Huxley, H.E. (1971) The structural basis of muscular contraction. *Proc. Roy. Soc. B*, **178**, 131–149.

Ingram, M. (1960) Bacterial multiplication in packed Wiltshire bacon. *J. Appl. Bact.*, **23**, 206–215.

International Commission on Microbiological Specification for Foods (ICMSF) (1988) *Microorganisms in Foods. Book 4. Application of Hazard Analysis Critical Control Point (HACCP) System in Ensuring Microbiological Safety and Quality*. Blackwell Scientific Publication, London.

Johnson, P. and Perry, S.V. (1970) Biological activity and the 3-methylhistidine content of actin and myosin. *Biochem. J.*, **119**, 293–298.

Jolley, P.D. (1979) The accumulation of unacceptably high levels of nitrite in vacuum-packed back bacon. *J. Food Technol.*, **14**, 81–87.

Katsaras, K. and Budras, K.-D. (1992). Microstructure of fermented sausage. *Meat Sci.*, **31**, 121–134.

Kitchell, A.G. and Ingram, M. (1963) Vacuum-packed sliced Wiltshire bacon. *Food Process Technol.*, **32**, 3–7.

Lawrie, R.A. (1952) Biochemical differences between red and white muscle. *Nature (London)*, **170**, 122–123.

Lawrie, R.A. (1953) The relation of energy-rich phosphate in muscle to myoglobin and to cytochrome oxidase activity. *Biochem. J.*, **55**, 305–309.

Lawrie, R.A. (1959) Water-binding capacity and drip formation in meat. *J. Refrig.*, **2**, 87–89.

Lawrie, R.A. (1966). Metabolic stresses which affect muscle. In *The Physiology and Biochemistry of Muscle as a Food*. Eds E.J. Briskey, R.G. Cassens and J.C. Trautman, pp. 137–144. University Wisconsin Press, Madison, WI.

Lawrie, R.A. (1978) Biochemistry of muscle in relation to growth. In *Patterns of Growth and Development in Cattle*. Eds H. de Boer and J. Martin, p. 65. Martinus Nijhoff, The Hague.

Lawrie, R.A. (1991) *Meat Science*, 5th edn., pp. 49, 61–81, 158, Pergamon Press, Oxford.

Lawrie, R.A., Pomeroy, R.W. and Williams, D.R. (1964) Studies on the muscles of meat animals. IV. Comparative composition of muscles from 'doppelender' and normal sibling heifers. *J. Agric. Sci.*, **62**, 89–92.

Lea, C.H. and Hannan, R.S. (1950) Biochemical and nutritional significance of reactions between proteins and reducing sugars. *Nature (London)*, **65**, 438–439.

Lea, C.H., Macfarlane, J.J. and Parr, L.J. (1960) Treatment of meat with ionizing radiations. IV. Radiation pasteurization of beef for chilled storage. *J. Sci. Food Agric.*, **11**, 690–694.

Ledward, D.A. (1981) Intermediate moisture meats. In *Developments in Meat Science. 2.* Ed. R.A. Lawrie, pp. 159–194. Applied Science Publishers, London.

Lefebvre, N., Thibault, C. and Charbonneau, R. (1992) Improvement of shelf-life and wholesomeness of ground beef by irradiation 1. Microbial aspects. *Meat Sci.*, **32**, 203–214.

Leistner, L.(1960) Microbiology of ham curing. *Proc. 12th Res. Conf. Am. Meat Inst. Fdn., Chicago, IL*, pp. 17–23.

Locker, R.H. (1987) The non-sliding filaments of the sarcomere. *Meat Sci.*, **20**, 217–236.

Locker, R.H. and Hagyard, C.J. (1963) A cold-shortening effect in beef muscles, *J. Sci. Food Agric.*, **14**, 787–793.

Love, S.B. and Haraldsson, S.B. (1961) *J. Sci. Food* Agric., **12**, 442.

McNeill, I.H., Dimick, P.J. and Mast, M.G. (1973) Use of chemical compounds and a rosemary spice extract in quality maintenance of deboned poultry meat, *J. Sci. Food Agric.*, **9**, 417–424.

Monin, G. and Sellier, p. (1985) Pork of low technological quality with a normal rate of muscle pH fall in the immediate *post mortem* period: the use of the Hampshire pig *Meat Sci.*, **13**, 49–63.

Mossel, D.A.A., Dijkmann, K.E. and Snyders, J.M.A. (1975) Microbial problems in handling and storage of fresh meats. In *Meat*. Eds D.J.A. Cole and R.A. Lawrie, pp. 223–246. Butterworth, London.

Murachi, T., Tanaka, K., Hatanaka, M. and Murakami, T. (1981) Intracellular Ca^{2+} dependent protease, (calpain) and its high molecular weight endogenous inhibitor (calpastatin). *Adv. Enzymol. Regul.*, **19**, 407–424.

Niven, C.F., Jr. (1958) Microbiological aspects of radiation preservation of food. *Annu. Rev. Microbiol.*, **12**, 507–524.

Niven, C.F., Jr. (1963) Technological aspects of radiation pasteurization of foods. *Int. J. Appl. Radn. Isotopes*, **14**, 26–29.

Ouali, A. and Talmant, A. (1990) Calpains and calpastatin distribution in bovine, porcine and ovine skeletal muscles. *Meat Sci.*, **28**, 331–348.

Ouali, A., Garrel, N., Obled, A., Deval, C., Valin, C. and Penny, I.F. (1987) Comparative action of cathepsins D, B, H and L, and of a new lysosomal cysteine proteinase in rabbit myofibrils. *Meat Sci.*, **19**, 83–100.

Paine, F.A. and Paine, H.Y. (1983) *A Handbook of Food Packaging.* Leonard Hill, Glasgow.

Paul, M.H. and Sperling, E. (1952) Cyclophorase system. XXIII. Correlation of cyclophorase activity and mitochondrial density in skeletal muscle. *Proc. Soc. Exp. Biol. Med.*, **79**, 352–354.

Penny, I.F. and Dransfield, E. (1979) Relationship between toughness and troponin T in conditioned beef. *Meat Sci.*, **3**, 135–142.

Penny, I.F., Voyle, C.A. and Lawrie, R.A. (1963) A comparison of freeze-dried beef muscles of high or low ultimate pH. *J. Sci. Food Agric.*, **14**, 535–543.

Raccach, M. and Henningsen, M. (1984) Role of lactic acid bacteria, curing salts, spices and temperature in controlling the growth of *Yersinia enterocolitica. J. Food Protect.*, **47**, 354–358.

Rhodes, D.N. and Shepherd, H.J. (1966) The treatment of meats with ionizing radiation. XIII. Pasteurization of beef and lamb. *J. Sci. Food Agric.*, **17**, 287–297.

Rolfe, E.J. (1958) *Fundamental Aspects of the Dehydration of Food.* Society of Chemical Industries, London.

Schaub, M.C. and Perry, S.V. (1969) The relaxing protein system of striated muscle. Resolution of the troponin complex into inhibitors and calcium ion sensitizing factors and their relationship to tropomyosin. *Biochem. J.*, **115**, 993–1004.

Schillinger, U., Kaya, M. and Lucke, F.K. (1991) Behaviour of *Listeria monocytogenes* in meat and its control by a bacteriocin-producing strain of *Lactobacillus sake. J. Appl. Bact.*, **70**, 473–474.

Schmidt, C.F., Lechowich, R.V. and Folinazzo, J.F. (1961) Growth and toxin production by type E *Cl. botulinum* below 40°F. *J. Food Sci.*, **26**, 626–630.

Scopes, R.K. (1964) The influence of *post mortem* conditions on the solubilities of sarcoplasmic proteins. *Biochem. J.*, **91**, 201–207.

Scopes, R.K. (1970) Characterization and studies of sarcoplasmic proteins. In *The Physiology and Biochemistry of Muscle as a Food. Vol. II.* Eds E.J. Briskey, R.G. Cassens and B.B. Marsh, pp. 471–492. University of Wisconsin Press, Madison, WI.

Scott, W.A. (1953) Water relations of *Staphylococcus aureus* at 30°C. *Aust. J. Biol. Sci.*, **6**, 549–564.

Scott, W.A. (1957) Water relations of food spoilage microorganisms. *Adv. Food Res.*, **7**, 88–127.

Scott, W.A. and Vickery, J.R. (1939) *Investigations on chilled beef. 2. Cooling and storage in the meatworks.* CSIRO Bull. No. 129, Melbourne.

Sharp, J.G. (1953) *Dehydrated Meat.* Spec. Ret. Food Invest. Bd., London, No. 57. HMSO, London.

Sharp, J.G. (1959) Observations on aseptic autolysis in muscle. *Proc. 5th-Meeting Eur. Meat Res. Workers, Paris*, paper 17.

Shaw, B.G. and Harding, C.D. (1984) A numerical taxonomic study of lactic acid bacteria from vacuum packed beef, pork, lamb and bacon. *J. Appl. Bact.*, **56**, 25–40.

Shaw, B.G. and Latty, J.B.A. (1984) A study of the relative incidence of different *Pseudomonas* groups in meat using a computer-assisted identification technique employing only carbon sources. *J. Appl. Bact.*, **57**, 59–67.

Shelef, L.A., Naglik, O.A. and Bogen, D.N. (1980) Sensitivity of some common bacteria to the spices sage, rosemary and allspice. *J. Food Sci.*, **45**, 1042–1044.

Sheridan, J.J., Lynch, B. and Harrington, D. (1992) The effect of boning and plant cleaning on the contamination of beef cuts in a commercial boning hall. *Meat Sci.*, **32**, 185–194.

Shorthose, W.R., Harris, P.V. and Bouton, P.E. (1972) The effects on some properties of beef of resting and feeding cattle after a long journey to slaughter. *Proc. Soc. Animal Prod.*, **9**, 387–391.

Smith, J.L. and Palumbo, S.A. (1981) Microorganisms as food additives. *J. Food Protect.*, **44**, 936–938.

Smith, M.G. and Graham, A. (1978) Distribution of *E. coli* and salmonellae on mutton carcass by treatment with hot water. *Meat Sci.*, **2**, 119–128.

Stolle, F.A. (1988) Establishing microbiological surveillance programmes at slaughterhouses. A new concept of meat hygiene. *Meat Sci.*, **22**, 203–212.

Takahashi, G., Lochner, J.V. and Marsh, B.B. (1984) Effects of low frequency electrical stimulation on beef tenderness. *Meat Sci.*, **11**, 207–225.

Tarr, H.L.A., Southcott, B.H. and Bissett, M.H. (1952) Experimental preservation of fish and beef using antibiotics. *Food Technol.*, **6**, 363, 366.

Tarrant, P.V. (1981) The occurrence, causes and economic consequences of dark-cutting beef: a survey of current information. In *The Problem of Dark Cutting in Beef.* Eds D.E. Hood and P.V. Tarrant, pp. 3–36. Martinus Nijhoff, The Hague.

Taylor, A.A. (1985) Packaging fresh meat. In *Developments in Meat Science. 3.* Ed. R.A. Lawrie, pp. 89–113. Elsevier Applied Science London.

Thornley, M.J., Ingram, M. and Barnes, E.M. (1960) The effects of antibiotics and irradiation on the *Pseudomonas–Achromobacter* flora of chilled poultry *J. Appl. Bact.* **23**, 487–498.

Waites, W.M. (1988) Meat microbiology: a reassessment. In *Developments in Meat Science. 4.* Ed. R.A. Lawrie, pp. 317–333. Elsevier Applied Science, London.

Walls, E.W. (1960) The microanatomy of muscle. In *The Structure and Function of Muscle.* Vol. 1. Ed. G.H. Bourne, pp. 21–61. Academic Press, New York.

Wang, K. and Williamson, C.L. (1980) Identification of the N_2 line protein of striated muscle. *Proc. Nat. Acad. Sci. USA*, **77**, 3254–3258.

Watts, B.M. (1954) Oxidative rancidity and discoloration in meat. *Adv. Food Res.*, **5**, 1–52.

Weber, H.H. and Meyer, K.H. (1933) Das kolloidale Verhälten der Muskelweisskörper. V. Das Mengenverhältnis der Muskelweisskörper in seiner Bedeutung für die Strucker des quergestreiften Kaninchenmuskels. *Biochem. Z.*, **266**, 137–147.

Weisser, H.H., Kunkle, L.E. and Deatherage, F.E. (1954) The use of antibiotics in meat processing. *Appl. Microbiol.*, **2**, 88–94.

Wierbicki, E. and Heligman, F. (1980) Irradiated bacon without and with reduced addition of nitrite. *Proc. 26th Meeting Eur. Meat Res. Workers, Colorado Springs*, **1**, 198–201.

Young, O.A., Torley, P.J. and Reid, D.H. (1992) Thermal scanning rheology of myofibrillar proteins from muscles of defined fibre type. *Meat Sci.*, **32**, 45–63.

2 Fermented meats — A world perspective

G. CAMPBELL-PLATT

2.1 Introduction

Fermented meat products are found in most parts of the world, although Europe is the major producer and consumer of these products.

Meat is a nutritious, protein-rich food which is highly perishable, with short shelf-life unless preservation methods are used. Fermentation is an important preservation method which has evolved for meat, but it is rarely used alone. Preservation is usually achieved by a combination of fermentation, with the use of water activity-lowering techniques including dehydration, and the addition of salt. These techniques have been the basis of traditional technologies used before the scientific basis of their action was understood.

Many of the salts used originally were not pure sodium chloride, but contained quantities of potassium or sodium nitrate. The nitrate helped preserve the pink meat colour and to direct the meat curing, initially by reduction to nitrite, so now nitrate or nitrite is often added directly to aid the fermentation and curing process. Spices are often used. As well as their role in providing desirable flavours in fermented meat products, they aid the safe fermentation by inhibiting undesirable or pathogenic micro-organisms. They can also have an important role in prolonging shelf-life of fermented meats through their antioxidant activity (Al-Jalay *et al.*, 1987). This is important as fat rancidity appears to be a limiting factor in dried sausage shelf-life at warm ambient temperatures (Seow and Goh, 1976).

With many fermented meat products, the selection of favourable conditions to encourage the growth and development of a desirable safe microflora is particularly important, as the meat may not be cooked, before or after the fermentation. In relatively few processes, especially among the traditional craft practices, are starter cultures used, and meat can be a good substrate for the growth of spoilage bacteria, such as *Pseudomonas*, or food poisoning bacteria including *Salmonella* and *Listeria* (Johnson *et al.*, 1990). As the meat surfaces dry, moulds can grow. This can be desirable and is encouraged as a part of the process, with *Penicillium nalgiovense* being of particular importance in meat fermentation. However, the conditions can also favour the establishment and growth of other undesirable fungi, several of which can produce potent mycotoxins.

Smoking can also be used to aid drying, and to deposit residues on the surface of the meat, which can inhibit surface mould growth. Alternatively, other antimould preservatives such as potassium sorbate can be used.

2.1.1 Types of fermented meat products

Fermented meat products can be divided into three groups, based upon the form of the meat.

1. Products made by using large, whole pieces of meat, with the country ham products forming a large part of this group. With some processes, the meat is cut into smaller strips or pieces, but still fermented as whole pieces. Examples include *biltong* and *jerky*.
2. The second, and major, group of fermented meats is made from meat which has been chopped into small pieces. These are the fermented sausages, with the various types of salami being well known.
3. The third, minor, group of fermented products is made from unusual parts of animals, such as bones or intestines. These products are produced in local areas, and are not well known in commercial products. Dirar (1992) has identified a number of such traditional products in Sudan.

The meats which are most commonly used are pork and beef. Lamb is also fermented, while chicken, duck and water buffalo are used in some products. Other types of meat are fermented occasionally in particular parts of the world, with gazelle, porcupine and whalemeat being three of the more unusual types.

Normally, the whole meat or meat pieces would be obtained from a single species, while the chopped fermented meat can originate from a mixture of species. An example of this is the Turkish sausage *soudjouk* which can be made from a combination of lean beef, water buffalo meat, mutton and sheep fat (Gokalp, 1986).

In Sudan, whole wild rabbit is skinned and then fermented for three days. The fermented carcass is then boiled in water, and pounded to give a thick meaty sauce (Dirar, 1992).

The actual significance of the fermentation depends on the extent to which other additives or processes are involved, the conditions, and the length of the process.

Many of these production methods have evolved over long periods from original craft processes. The science behind the processes may have been little studied, particularly in the case of many traditional fermented meat products. In other cases, where the scale of production has been increased and production has been concentrated in larger factories, technology has been applied. This has often been applied to ways of accelerating the production process, as in the injection brining of hams, in which the role of fermentation is reduced. Often these newer, lower-cost products have become popular. Sweetcure, higher moisture bacon and hams need more refrigeration or sophisticated packaging. Generally, there has been a move towards the milder-flavoured, less salty meat products. Counterbalancing this to some extent, as products have become more similar, has been a greater interest in eating lesser-known foods from different parts of the world. It is here that there could be continued or renewed interest in some of the lesser-known, more traditionally fermented meat products.

2.2   Fermented whole-meat products

The major fermented whole-meat product is country ham, or country-style ham, which is made from pork. These hams are produced principally in Europe and North America with some production in South America, Asia, Oceania and Africa. Several types are listed in Table 2.1.

Because the pork is not usually cooked, it should be examined and found free from parasites, particularly *Trichinella spiralis*, the pork tapeworm. The meat is prepared by removing the aitch bone and shank, and then, in the traditional, fermentation process, the meat surface is rubbed with a dry mixture of sodium chloride and potassium or sodium nitrate. Spices may be added as flavourings, and may include pepper, allspice, coriander, mustard or juniper berries. A small quantity of sugar or molasses is sometimes added. The curing process takes weeks or months.

The initial salting is done at low temperature, normally 5–10°C for 10–15 days. Then further salt pickle is usually added, and this continues for a further 20–50 days. During the drying–maturation stage, the temperature is increased in stages over several weeks or months. Temperatures during these stages are increased typically from 13–18°C to around 30–35°C, or ambient temperatures in the later stages of drying. Cool smoking at 30–40°C is done over a period of days or weeks with some of the hams. Under these high-salt and nitrate conditions, coagulase-negative micrococci and staphylococci bacteria tend to dominate the microflora. It is important to use high standards of hygiene, because enterotoxinogenic strains of contaminating *Staphylococcus aureus* have been isolated from dry-cured hams (Marin *et al.*, 1992) (Table 2.1).

There are a number of semi-dried uncooked meat products in which fermentation plays a minor part along with chemical curing. The meat, typically beef, but may be sheep, goat or other meat, is cut into pieces of 2–10 kg which are then heavily salted with sodium chloride, nitrate or nitrite, spices, and seasonings added. Bacteria of *Micrococcus* or *Staphylococcus* spp. reduce nitrate to nitrite, while lactic acid bacteria help reduce the pH. Yeasts, often of *Debaryomyces* spp. may be active, helping with flavour development. Drying is encouraged by using low temperatures, 5–15°C, and high humidities, 80–90%, initially, followed by lower humidity

Table 2.1 Some types of country ham

Type	Area of production	Characteristics
Europe		
Ardennes	Belgium	–
Bayonne	Basque, France, Spain	Rosemary, red wine and olive oil added to brine. Lightly smoked
Belfast	Northern Ireland	Smoked, usually over peat
Bradenham or Suffolk	Suffolk, England	Molasses gives black skin, smoked
Coppa	Parma, Italy	–
Cumberland	Northern England	–
Iberico	Spain	–
Jambon Blanc	France	Hot smoked
Jambon de Paris	France	Hot smoked
Kasseler	Germany, Netherlands, Scandinavia	Hot smoked over wood and juniper berries
Katenspeck	Germany, Austria, Switzerland	Belly of pork, hot smoked, black surface
Kraski Prsut	Yugoslavia	–
Lachsschinken	Germany	Pork loin, wrapped in fat, smoked
Limerick	Ireland	Smoked over peat
Parma	Parma, Italy	Pepper, allspice, coriander and mustard seasoning
Prosciutto	Italy	–
Salt Pork	–	–
San Daniele	Italy	–
Schwarzwalder	Black Forest, Germany	Smoked
Scotch	Scotland	–
Serrano	Spain	–
Spekeskinke	Norway	–
Suffolk or Bradenham	Suffolk, England	Soaked in molasses
Westphalian	Westphalia, Germany	Smoked over juniper on beechwood
York	Northern England	Hot smoked
North America		
Kentucky	Kentucky, USA	–
Salt Pork	USA	–
Scotch	USA	–
Smithfield	Virginia, USA	Smoked with hickory, deep red colour
Asia, East and South-East		
Ching Hua	China	Red colour
Mu-uan	Southern Thailand	–
Yunnan	China	Red colour

as drying proceeds. Cold smoking at 30–40°C may be used over several days, followed by a maturation period of several weeks or months. The high-salt content and drying reduce the water activity (a_w) to below 0.9 in the higher-water content products down to 0.75 in the lower-water content, more stable meats.

European products of this type made from beef include *bresaola* in Italy, and *Bundnerfleisch* a smoked product from Switzerland. *Pastrami*, which may be made from beef, goat or mutton, spices with nutmeg, paprika, allspice and garlic, then smoked, originated in Romania, but is also made in the USA. *Fenelar* is smoked mutton from Norway. Salt beef is made in Israel and the USA, while other salt meats are found in Europe and the Middle East.

In Central and South America, *carne de sol* is made in Brazil, while *carne seca* is made in Mexico and many countries of South America. Both of these products are usually made from beef.

In Asia, some similar products are found. These include *nuakhem*, made from beef in Thailand, and *bebontot*, a traditional product from Bali, Indonesia. To make *bebontot*, meat, usually pork but sometimes beef, is cut into cubes of about 20 mm, then mixed with turmeric, galangal, ginger, garlic, chillies and pepper, before drying in palm sheaths. Also in Indonesia, *dendeng* is made by adding sugar, salt and spices to meat, which is then sun-dried for 4–5 days. It is normally eaten after deep-frying in oil, and eaten with rice (Muchtadi, 1986).

In Malaysia, *serunding* is a ready-to-eat low-moisture meat product of two types: *serunding daging* is made from beef, while *serunding ayam* is made from chicken (Lim Chin Lam, 1989). These are shredded spiced meats, in which salt and sugar are used, together with coconut milk and a mixture of spices including ginger, chilli, pepper, coriander, garlic, shallot and Asam gelugur. Usually after a short pickling or fermentation period, the product is cooked and then canned and given a second cooking in a retort. In this respect it is similar to corned beef which is produced in Europe, North and South America and consumed worldwide, as a canned ready-to-eat product which stores and travels well.

In hotter regions of the world, it has been found necessary to cut meat into thin strips or slices to achieve rapid drying before putrefaction. Most of these products use lean meat, which when dried and fermented produce useful high-protein reserve food which can be used by travellers, eaten at times of celebration, or can be added as a flavoursome, nutritious supplement to soups and stews. In these products, bacteria of the genera *Micrococcus* and *Staphylococcus* often initiate reduction of nitrate to nitrite, with some growth of lactic acid bacteria (Prior, 1984). The combined action of salt and dehydration is the main form of preservation, with the product having final a_w of 0.73–0.82, and a water content of 10–25%.

Mould growth can be a problem in prolonged storage of these products at high ambient temperatures. Although the use of lean meat normally gives a low-fat content of 2–5 g per 100 g dry matter, some of these products are sealed and stored in fat, which helps exclude oxygen and inhibits mould growth. Examples of the use of fat in this way include *pemmican* used originally by the Indians of North America to preserve beef, buffalo or reindeer. In Libya, lamb or beef is sun-dried for 2–4 days by most households to produce *giddeed*. This can then be deep fried in olive oil to produce *gergush*, which is sealed in glass jars, and is kept and used over several months.

There are many of these products made in different parts of Africa. In Southern Africa, particularly South Africa, Lesotho and Zimbabwe, beef or antelope is dried in long strips to produce *biltong*. In north-east Africa, *shermute* or *shermout* is made from beef in Sudan, while in West Africa, antelope or other game meat is dried into *mpu nam* in Ghana, and *bunda* in Nigeria. Also in Nigeria, and in Niger, *kilishi* is produced. *Kilishi* is usually made from muscles of hindquarters of beef, mutton or goat (Igene *et al.*, 1992). The slices of meat are soaked in a pickle mixture containing salt, spices, sugar and other ingredients including defatted groundnut paste (*Arachis hypogaea*) and pepper. The meat is then sun-dried, and then given a rapid roasting. The final product contains some 50% protein, and has a final water content of about 7%. Because of its stability at high ambient temperatures, it is now produced by traditional meat processors in commercial quantities in northern Nigeria, where it is widely available; it is exported to Saudi Arabia, particularly at times of annual pilgrimage there.

Kundi is produced in Nigeria by smoke-drying strip pieces of meat, usually beef. The smoking method using high temperatures has been found to result in high levels of carcinogenic benzopyrene and other polycyclic aromatic hydrocarbons (Alonge, 1988).

In South America, *charqui* or *sharqui* is made in flat sheets from beef, or sheep in Brazil, or may also be made from llama or alpaca in the higher regions of Peru. Jerked beef or *jerky* are produced in South America and in the Caribbean region, while *jerk* pork is a favourite speciality in Jamaica.

Iqunaq is an unusual product from the world's coldest region. It is made by the Inuits of Greenland and northern Canada. It is made from the eider duck. The ducks are strangled, then put in layers between stones and allowed to ferment. The fermentation appears to be due to a mixture of microbial and autolytic enzymes. The relatively high pH of the product has made it an occasional source of botulism poisoning from the growth of *Clostridium botulinum* when the fermentation is slow or incomplete (Borgstrom, 1968).

2.3 Chopped or comminuted fermented meats: sausage

There is a large number of these products made from chopped, comminuted meat, with or without additives. The term 'sausage' does not mean that the products are necessarily fermented. There are number of fresh, non-fermented sausages, which must always be cooked before eating. In the UK, most sausages are of this type. The well-known *Frankfurter* which originated in Frankfurt, Germany and Vienna, Austria, and is now found throughout Europe, North America and most parts of the world, is a cured sausage which is usually eaten hot.

While fermented sausage production is now found in most parts of the world, Europe has long dominated production. It is estimated that greater than one million tonnes of fermented sausage is produced in Europe annually, with Germans being the largest consumers, eating on average about 5 kg per person annually (Lücke, 1985).

Raw meat, most often pork or beef, is chopped into small pieces, and may be mixed with various other ingredients before being stuffed into casings, which were originally animal intestines. The additives would usually include sodium chloride, sometimes nitrate or nitrites, occasionally a small amount of sugar, plus various spices and seasonings.

To direct these sausage fermentations 'back-slopping' was used originally, in which some meat from a previous successful fermentation was added to encourage establishment of the desired microflora. Nowadays, many manufacturers use starter cultures (Smith and Palumbo, 1983). These consist of a single culture or mixture of lactic acid bacteria of *Pediococcus* spp., *Lactobacillus plantarum* or *Lactobacillus brevis*. *Micrococcus* or non-pathogenic *Staphylococcus* spp. may also be used, and are helpful in reducing nitrate to nitrite.

These nitrate-reducing micrococci and coagulase-negative staphylococci also produce catalase which removes the hydrogen peroxide produced by the lactic acid bacteria. Typically, with the lactic acid bacteria producing lactic and other organic acids, the pH value of the meat falls from 5.8 to around 4.8 within one month in German salami production. With some fermented sausage, as in Italy, mould grows on the surface and reverses the fall in pH, giving a final pH value of 6.0–6.2 (Marchesini *et al.*, 1992).

Preservation is achieved by a combination of fermentation through microbial action, and lowering of water activity through addition of salt and dehydration. With some products, smoking further aids preservation, and particularly inhibits surface mould growth.

Hygiene and preservation is very important in fermented sausage production. In recent times, as the potential seriousness of the presence of *Listeria monocytogenes* became apparent, surveys of fermented sausage products have found the organism to be present in 5–32% of the product (Varabioff, 1992). Whilst during fermentation of sausages, *L. monocytogenes* has been found to decrease substantially, the organism is psychotrophic, grows in up to 10% sodium chloride and in the presence of nitrite, and is therefore of major concern. Pediocin, the bacteriocin produced by some strains of *Pediococcus acidilactici* in sausage fermentation, has been found to inhibit *L. monocytogenes* during sausage drying (Foegeding *et al.*, 1992).

The length of process involved and the shelf stability of the products depend on the final water content achieved from an initial water content of meat of approximately 75%.

The sausages can therefore be conveniently classified by water content into:

Moist:	50–60%
Semi-dried:	35–50%
Dried:	20–35%

2.3.1 Moist sausages

Because of the high moisture content, these products are highly perishable, and, unless further preservation methods have been applied, should be refrigerated and used rapidly. In Germany, the finely-chopped moist spreadable sausages *Mettwurst* and *Teewurst* are popular, but should be consumed within one or two days.

Merguez-type moist-fermented sausages are produced in warmer countries of southern Europe and North Africa, particularly Algeria, Tunisia, Morocco and southern France. *Merguez*-type sausages are perishable, and are cooked, usually fried, before eating. They are made from pork, beef, lamb or mixed meats. The meat is coarse-chopped, mixed with sodium chloride, potassium nitrate or sodium nitrite, plus often sodium ascorbate and seasonings, then chopped with fat and filled into medium-large casings of diameter 50–70 mm. Initial mixing and filling is at temperatures around 0°C. This is followed by a high-temperature, short-time fermentation of 30–35°C over 3–6 days. The temperature and humidity are then reduced during a brief drying period of a few days.

Another type of moist sausage comes from Lebanon, Pennsylvania, in the USA. The *Lebanon Bologna* is a coarse-chopped beef sausage which is heavily smoked (Palumbo *et al.*, 1973). The beef is aged for several days in 2–3% sodium chloride, which selectively encourages the growth of micrococci. This helps the rapid conversion of 400–600 mg/kg added potassium nitrate, to nitrite, which, together with added sugar, encourages a lactic fermentation at the high temperature of 35–43°C which is used. During this period under high humidity, the meat is smoked, and then the sausage is cooled at 4–6°C for three days. The high-temperature fermentation allows unusually rapid *Lactobacillus* and *Lactococcus* growth, giving high lactic acid yields and a consequent low pH value of 4.3–4.9.

Mortadella is made in North America and also in Europe in Italy and France. It is made from pork, and lactose may be added as a carbohydrate source from milk powder. It is normally not smoked. Like *Lebanon Bologna*, it is subjected to a high-temperature fermentation when *Pediococcus* and *Lactobacillus* species produce significant levels of lactic acid, giving a low pH. After production, the sausage, in its larger diameter casings, is cooked to give an internal pasteurisation temperature of 65–70°C. Some drying, at lower humidity, is then applied.

Kochsalami is another similar product made in Germany and the USA. It is made from coarse-chopped beef, and/or pork, and is heavily seasoned. After a high-temperature fermentation, it is smoked and cooked before a partial drying stage.

2.3.2 Semi-dry sausages

Several types of these fermented sausage originated in Europe, but their popularity spread, with emigrants, across the Atlantic to the USA. These semi-dry *summer sausage* or *farmer sausage* now dominate production of fermented sausage in the USA.

Pork or beef is usually used, sometimes mixed together. Sodium chloride is added, together with potassium nitrate or sodium nitrite as curing salts, plus sugar and seasoning. The addition of pork fat is done at low temperature, −2 to −4°C, to prevent smearing, and then the meats are stuffed into casings. Fermentation is carried out at higher temperature and humidity typically 25–32°C at 85–95% relative humidity. After the initial fermentation, smoking is normally applied at 35–50°C for 12–48 hours. The temperature is then reduced to 14–18°C, to allow a ripening or maturation period over 10–25 days under reducing humidity. During this period, shrinkage occurs and flavour develops from microbial and endogenous enzyme activity.

If nitrate is used, *Micrococcus* and *Staphylococcus* bacteria play an important role in its reduction to nitrite. During the fermentation, lactic acid bacteria, principally belonging to *Pediococcus acidilactici* and *Lactobacillus plantarum*, are the most active. Particularly in the USA, increasing use is made of added starter cultures. These normally contain mesophilic *Pediococcus* spp. and *Lactobacillus plantarum* (Hammes *et al.*, 1990).

An unusual variant is *isterband* which is made in Sweden and contains barley grains in addition to beef and pork meats.

In France and Portugal, *longaniza* is produced, which is a beef or pork sausage, heavily seasoned with aniseed (*Pimpinella anisum*) and pimento and allspice (*Pimenta officinalis*).

Figatelli is made in France from pork with some pork liver mixed in. It is often fried before eating.

Whilst there is a long history of semi-dried fermented sausage production in Europe, there is an equally long history, dating back over 2000 years, of the production of related products in South-East Asia and China (Anderson, 1988).

China is the world's largest producer of pigs, which are its main source of meat. Chinese sausage, or *laap ch'eung*, is therefore made from pork, which is usually coarsely chopped, and sometimes pork liver may be added (Anderson and Anderson, 1977). Sodium chloride and sugar are added, plus sometimes cooked rice as a further carbohydrate source. Garlic and pepper is often used as seasoning. The sausages are generally packed in thin casings (25–30 mm diameter) which allows rapid drying as the temperature is raised to 35–45°C for 2–3 days. The temperature is then reduced to 20–25°C to allow further ripening.

The Chinese sausage *xun chang* is made from chopped pork, which is spiced and filled into pig intestines. These are smoked, normally over pinewood, and then finally sun-dried. They are often eaten after cooking in boiling water or steaming for 15–20 minutes. These are made traditionally in households, particularly at the time of the Chinese New Year spring festival.

Production of similar fermented products is found throughout the region, including in Vietnam, Laos and northern Thailand. These products are local variations of the process probably originating in China, but sometimes the meat is wrapped in packets made from banana-leaves, before fermenting at the high ambient temperature of 25–35°C for 3–5 days. In Thailand the varieties of sausage are known as *naam* or *naem*, or *musom*. A further variety, *sai-krok-pries*, is produced in north-eastern Thailand from pork or beef (Sundhagul *et al.*, 1975). Normally, all these Chinese and Asian sausages are cooked by steaming before eating, which provides an extra safety hurdle.

2.3.3 Dried sausages

These are the longer-life sausages which originated in Europe as a way of preserving meat throughout the summer months before refrigeration became available. The particular acidic, seasoning, smoked and other characteristic flavours produced during the long fermentation and ripening period of several weeks, created a following for these products. As European peoples emigrated to North America, South America, Southern Africa or Australia they took with them a knowledge and taste for these products, which are now produced in significant quantities in many parts of the world. One particularly unusual example of this transfer of technology is the production of *chourisam*, a heavily-seasoned dried pork with pork liver sausage made in Goa on the west coast of India which was settled by the Portuguese. It is based upon *chorizo*, the main spiced and coloured pork or beef sausage from Spain, and which may be fried before eating.

Another variant, *chorizo Pamplona*, made from a mixture of coarse-chopped beef and pork, and containing paprika, travelled from Spain westwards and is now a Mexican dried fermented sausage.

A further variety, *Veneto salami*, is made in Australia from medium-chopped beef, with crushed peppercorns and red wine. This inclusion of an alcoholic beverage in fermented sausage is unusual, but not unique because Anderson (1988) reported on a sausage prepared in Canton, China containing a distilled rose-flavoured liquor.

Pepperoni is another heavily-seasoned dried fermented sausage, which is smoked. It is usually made from pork or beef, which may be mixed and other meats, such as lamb may be included. Relatively small diameter casings (15–40 mm) are normally used, and moulds of *Penicillium* and *Aspergillus* spp. may contribute to flavour by growing on the surface during maturation. Most are produced in central, northern and eastern Europe, with production also in North America and Australia. The strong flavour of pepperoni-type sausages has made them, when sliced, a popular ingredient of the topping of pizza. Slices are also used in soups and stews particularly in central and eastern Europe.

Particular variants, in which different meats are used, include *fjellmorr gilde* in Norway, and *lambaspaeipylsa* in Iceland which include lamb in addition to beef and pork. Horse meat may also be added to *toppen* and *trøndermorr* in Norway.

In many of the same European countries pork and beef are medium-chopped, mixed together and filled into medium-sized casings (40–65 mm) in the production of seasoned, normally smoked, German or Hungarian-style salami.

Many types of salami are produced and a large number are listed in Campbell-Platt (1987). Whilst most are based on pork or beef, lamb is sometimes incorporated with the pork or beef, particularly in Norway. Goat and horse meat can be added to *farepølse* in Norway, while horse meat may be added to *sognekorr* or *sognemorr gilde*, and *stabbur* in Norway and *kotimainen meetwurst* in Finland. Reindeer meat may also be incorporated in Scandinavian countries; examples include *poro meetwurst* in Finland and *rallersnabb gilde* and *reinsdyr pølse* in Norway.

In the production of Italian-type salami, the same cooler fermentation temperature of 15–25° C is used, followed by a prolonged drying period at 12–15° C, when relative humidity is progressively decreased. The final water activity ranges from 0.67 to 0.92, allowing mould growth (Holley, 1981). These type of salamis are not smoked, and fungi may colonise their surface during drying. As with German salami, increasing use is being made of starter cultures, but in the case of Italian-type salami, the starter culture may not contain just lactic acid bacteria, but also the yeast *Debaryomyces hansenii* (Coretti, 1977). By dipping the sausage in a suspension of conidia of *Penicillium nalgiovense*, the growth of undesirable, probably mycotoxinogenic moulds can be inhibited.

An interesting example of how Italian salami production has moved around the world is illustrated by *Milano*, a dried sausage made of beef, or beef and pork, usually with garlic which originated from Milan, Italy. It is now found also being produced in Greece, Australia, USA, Mexico,

Uruguay and Brazil. Similarly, *cacciaturi* and *varzi* have both moved from Italy to Mexico.

2.4 Fermented products made from offal and other parts of animals

While sausages are often made from varying proportions of non-muscle meat trimmings and offal, there are some unusual products which are made wholly from non-muscle parts of the animal.

As with other forms of 'fermented meat', other forms of preservation are involved and fermentation is only part of the preservation or processing method used.

A number of these unusual products are found in north-east Africa, particularly Sudan, and have been described by Dirar (1992). *Miriss* is made from the peritoneal fat surrounding lambs' stomachs, and *mussran* is made from small intestines. In both, alkaline potash from plant material ash is mixed in, and a proteolytic fermentation takes place in an earthenware vessel. After three days, *mussran* is removed for sun drying. The *miriss* fermentation continues for about six days to give a strong foul-smelling white product.

Intestines are also used to make *twini-digla*. They are cut into long strips which are left in the sun. Long strips of other internal organs, such as the heart, liver, kidneys and spleen are also sun-dried, and then all are pounded together. Salt, and ash, probably containing some nitrate, is added and the resulting paste is fermented and sun-dried for a further week. *Um-tibay* is made by fermenting a similar range of ingredients in an empty rumen casing. Traditionally this is fermented at the top of a high support. The strong-smelling product is then cooked in hot sand and fire embers overnight, before direct consumption, or after further drying.

Beirta is made by fermenting a mixture of goat meat with offal, and some milk added. After about four days, salt is added and a further three days fermentation continues in an earthenware vessel.

Also in Sudan, two unusual products are made from crushed animal bones. *Kaidu-digla* is made by fermenting and sun-drying chopped bones mixed with water and ash for 2–3 weeks. The dry balls are stored for later used, pounded into soups and sauces.

Dodery, mulaa el-sebit or *aki-el-muluk* is made from crushed bones, marrow and fat. After initial fermentation in an earthenware pot for about three days, ash is mixed in and further fermented for another five days. The resultant product is a strong-smelling pungent brown paste, which is used particularly at festival time.

Many of these products may seem unattractive to outsiders, but they are rooted in the traditions of cultures in particular regions. They represent ways of preserving and utilising all parts of the valuable animals. Particularly where extensive proteolytic breakdown occurs, the products have pungent odours, which make them unattractive to people who are not used to consuming them.

It seems unlikely that these local products will therefore develop beyond their local areas of production and consumption. The role that fermentation plays and the nature of the changes taking place during their production, is poorly understood.

Other unusual local fermented meat products from Africa and other regions of the world are yet to be described.

2.5 Conclusion

Worldwide, the fermented sausages and fermented whole meat products play an important part in many diets. They provide valuable protein and fat nutrients, and interesting flavours. These products extend considerably the life of perishable meat, from days to weeks and months, even at warmer temperatures. This preservation is achieved partly by fermentation and partly by dehydration and lowering of water activity. Although preserved, these nutrient-rich products can be hazardous, and the trend towards milder processing means that good hygienic practices, better control and the use of refrigeration become more important in ensuring that these products can continue to be both enjoyable and safe.

References

Al-Jalay, B., Black, G., McConnell, B. and Al-Khayat, M. (1987) Antioxidant activity of selected spices used in fermented meat sausage. *J. Food Protect*, **50**, 25–27.

Alonge, D.O. (1988) Carcinogenic polycyclic aromatic hydrocarbons (PAH) determined in Nigerian *kundi* (smoke-dried meat). *J. Sci. Food Agric.*, **43**, 167–172.

Anderson, E.N. Jr. (1988) *The Food of China*. Yale University Press, New Haven CT.

Anderson, E.N. Jr. and Anderson, M.L. (1977) Modern China: South. In *Food in Chinese Culture*. Ed. K.C. Chang pp. 318–382 Yale University Press, New Haven CT.

Borgstrom, G. (1968) *Principles of Food Science*. Vol. 2. *Food Microbiology and Biochemistry*, pp. 103–126. Macmillan, New York.

Campbell-Platt, G. (1987) *Fermented Foods of the World — A* Dictionary and Guide. 314 pp. Butterworths, London.

Coretti, K. (1977) Starterkulturen in der Fleischwirtschaft. *Fleischwirtschaft*, **57**, 386–394.

Dirar, H.A. (1992) The indigenous fermented foods and beverages of Sudan. In *Applications of Biotechnology in Food Processing in Africa*, pp. 23–40 UNIDO, Vienna.

Foegeding, P.M., Thomas, A.B., Pilkington, D.H. and Klaenhammer, T.R. (1992) Enhanced control of *Listeria monocytogenes* by *in-situ*-produced pediocin during dry-fermented sausage production. *Appl. Environ. Microbiol.*, **58**, 884–890.

Gökalp, H.Y. (1986) Residual nitrate, nitrite, carbonyl and TBA values of Turkish *soudjouk* manufactured by adding different starter cultures and using different ripening temperatures. *J. Food Technol.*, **21**, 615–625.

Hammes, W.P., Bantleon, A. and Muis, S. (1990) Lactic acid bacteria in meat fermentation. *FEMS Microbiol. Rev.*, **87**, 165–174.

Holley, R.A. (1981) Prevention of surface mould growth on Italian dry sausage by natamycin and potassium sorbate. *Appl. Environ. Microbiol.*, **41**, 422–429.

Igene, J.O., Farouk, M.M. and Akanbi, C.T. (1992) Preliminary studies on the traditional processing of *kilishi*. *J. Sci. Food Agric.*, **50**, 89–98.

Johnson, J.L., Doyle, M.P. and Carrens, R.G. (1990) *Listeria monocytogenes* and other *Listeria* spp, in meat and meat products. A review. *J. Food Protect.* **53**, 81–91.

Lim Chin Lam (1989) Traditional Malay foods: A review. *Bull. Res. Inst. Food Sci. Kyoto Univ.*, 28–73.

Lücke, F.K. (1985) Fermented sausages. In *Microbiology of Fermented Foods*. Ed. B.J.B. Wood, Vol. 2, pp. 41–83. Elsevier Applied Science, London.

Marchesini, B., Bruttin, A., Romailler, N., Moreton, R.S., Stucchi, C. and Sozzi, T. (1992) Microbiological events during commercial meat fermentations. *J. Appl. Bact.* **73**, 203–209.

Marin, M.E., de la Rosa, M.D.C. and Cornejo, I. (1992) Enterotoxinogenicity of *Staphylococcus* strains isolated from Spanish dry-cured hams. *Appl. Environ. Microbiol.*, **58**, 1067–1069.

Muchtadi, D. (1986) *Dendeng*, a cheap method of meat preservation — effective but destructive. *Food Lab. News*, **5**, 19.

Palumbo, S.A., Smith, J.L. and Ackerman, S.A. (1973) Lebanon Bologna. I. Manufacture and processing. *J. Milk Food Tech.* **36**, 492–503.

Prior, B.A (1984) Role of microorganisms in *biltong* flavour development. *J. Appl. Bact.*, **56**, 41–45.

Seow, C.C. and Goh, S.A. (1976) Lipid oxidation in the Chinese sausage. *Mal. Agric. Res.*, **5**, 163–169.

Smith, J.L. and Palumbo, S.A. (1983) Use of starter cultures in meats. *J. Food Protect.*, **46**, 997–1006.

Sundhagul, M., Daengsubha, W. and Suyanandana, P. (1975) Thailand's traditional fermented food products: a brief description. *Thai J. Agric. Sci.*, **8**, 205–219.

Varabioff, Y. (1992) Incidence of *Listeria* in smallgoods. *Lett. Appl. Microbiol.*, **14**, 167–169.

3 Historical aspects of meat fermentations

P. ZEUTHEN

3.1 Introduction

Most reviews on fermented meats point out that drying and fermentation are probably the oldest forms of preservation (e.g. Bacus, 1984; Smith, 1987; Roca and Incze, 1990). The two processes are mentioned together, because in practice they are impossible to separate. To these may be added smoking, which undoubtedly is at least as old a preservation method as drying. The authors quoted above claim that these preservation methods are several thousands of years old. Smith (1987), thus makes reference to Homer's *Odyssey*, ca. 900 BC, and sausages of the old Roman Empire. Leistner (1986a) citing Lissner (1939) mentions that 'sausage', as such, is an ancient word in many languages. Thus, *Wurst* is an Indo-Germanic word, probably derived from Latin, meaning 'to turn' or 'to twist'. Sausage is also well-known as *Kolbasa* in Slavic, derived from Hebrew, meaning 'all kinds of meats'. The origin of the Danish or rather Scandinavian word *pølse* is uncertain, but is probably derived from the Latin word *pulvinus*, meaning a 'cylindrical pillow'. Similarly, the origin of the word *salami* seems uncertain. Most authors, such as Leistner (1986a) say that it is derived from Latin, simply meaning 'salt', whereas Bacus (1984) claims that it is derived from the name of the city Salamis on Cyprus. Most authors (e.g. Adams, 1986) claim that the production of fermented sausages is thought to have originated in the countries surrounding the Mediterranean Sea, although he also mentions *Nham*, the fermented pork sausage of Thailand as an example of other areas, where fermentation of meat was of local origin. Leistner (1986a) mentions that Chinese sausages were known in the Chinese North and South Dynasties, 589–420 BC, although these sausages were far from fermented or non-fermented sausages as we know them; besides, they were intended to be consumed hot. It seems that the art of sausage production, in most cases, can be traced back to southern Europe, where it first spread to other European countries. Leistner (1986a) thus mentions that the most well-known cured and fermented German sausages, probably first produced by Italians, are only ca. 250 years old, and the Hungarian salami is no more than 150 years old. Adams (1986) writes that the European emigrants established production in both the USA, South America and Australia, and that knowledge about fermented sausages in the Seychelles, the Philippines and Papua New Guinea is largely a result of European influences, although food fermentation as such is well known. As late as 1971, the manufacture of fermented sausage was described as 'an art' (Kramlich, 1971).

Climatic variations have of course also influenced the ability of people to produce dried — and fermented — meat products. Low relative humidity has always been an indispensable requirement for subsequent drying and 'finishing', a condition which can be found in mountainous regions in Spain and Italy, but also in the Ardennes. In more modern times, up to a few decades ago, it was well known that it was impossible to produce fermented meats in many valleys and other fairly humid places without the use of mechanical refrigeration and drying facilities. The author knows of one large Danish meat manufacturing plant where this was the case, although drying and production of fermented sausages in Denmark has been common for a very long time. The climatic variations during the year have even influenced the names of fermented sausages in some countries. Thus Incze (1986) writes that the Hungarian salami, the so-called 'winter salami' originated in Italy, where the climate in northern Italy was far better suited for the production of fermented sausages, with fairly low temperatures and relative humidities. Although the conditions were less optimal in Hungary, it turned out to be possible to dry fermented sausages in Hungary without major difficulties, in particular during the winter months; hence the name 'winter salami'. Similarly, the name 'summer sausage' was given because the sausage was mainly manufactured during the summer, where for safety and shelf-life reasons microbial growth was stopped by heating the sausage as the last processing step.

3.2 Terminology of fermented meats and names

There are a number of opinions about the origin of meat fermentations. No doubt, early manufacturers did not originally realise that in drying meat and meat products more or less quickly (and successfully), they actually subjected the products to some degree of fermentation. Lücke (1985) thus maintains that in Germany, the manufacture of fermented sausages only commenced some 150 years ago, and most of the fermented sausages were smoked. Other types of fermented sausages emerged later as a consequence of advanced meat processing techniques and the availability of refrigeration. These products include spreadable, non-dried or semi-dry sausages. Other criteria in his review include the casing diameter, degree of comminution of the ingredients, animal species and seasoning. He also uses the term 'sausage ripening' to describe changes occurring between casing filling and the time when the product is ready for sale, while the term 'sausage fermentation' covers the actual lactic acid formation and concomitant processes. Based on these criteria Lücke (1985) classified

Table 3.1 Classification of fermented sausages (from Lücke, 1985)

Product type	Weight loss during drying*	Smoked	Growth of moulds and yeasts	Examples
Dry sausage air-dried	>30%	no**	yes	Genuine; salami, French saucisson
Smoked	>20%	yes	no	German Katenrauchwurst
Semi-dry sausage	<20%	yes	no	Summer sausage
Undried spreadable fermented sausage	<10%	usually	no	German Teewurst, frische Mettwurst

* Approximate.
** Or only lightly smoked during fermentation.

fermented sausages as shown in Table 3.1. Lücke finally explains that in his opinion, intensive research in sausage fermentations was not initiated until the traditional empirical methods of manufacture no longer met the requirements of large-scale, low-cost industrial production with short ripening times and highly standardised products, i.e. in the USA in the 1930s and in Europe in the 1950s, when the first systematic studies on the microbiology and production of fermented sausage were first published. In his book, Coretti (1971) also defines fermentation as the production of lactic acid, but he uses the term 'ripening' to refer to all the chemical, physical, microbiological and enzymic changes taking place in the sausage and which are temperature and humidity controlled. In his book, Kinsman (1980) classifies sausages into six categories:

1. *Fresh sausages*. Made of chopped (comminuted) meats that are neither cured or smoked. Seasoned and usually stuffed into casings but may be prepared in bulk form. This type of sausage should be fully cooked before serving. Examples include *Bockwurst*, country-style pork sausage, fresh pork sausage, Italian-style sausage and pork sausage roll.
2. *Cooked sausages*. Made from non-cured meats, ground, seasoned, stuffed into casings, and cooked, but not smoked. Usually served cold. Examples include Braunschweiger liver sausage and *Leberkäse* (liver cheese).
3. *Cooked smoked sausages*. Made from cured meats, chopped or ground, seasoned, stuffed into casings, smoked lightly and then fully cooked. The types do not require further cooking before consuming but are often heated before serving. Examples include Berliner, Bologna, *cotto salami*, frankfurters, smoke links and Wieners.
4. *Uncooked smoked sausages*. Made from either cured or uncured meats, comminuted, seasoned, stuffed and smoked, but not cooked. These must be fully cooked before consuming. Examples include Kielbasa, Mettwurst and smoked country-style pork sausage.

5. *Dry and/or semi-dry sausages.* Sometimes called 'summer sausage', made from cured meats which are comminuted, seasoned, stuffed and air dried under controlled time–temperature–humidity conditions. These may or may not be smoked before drying. Examples include *cappicola*, *chorizos*, farmer *cervelat*, frizzes, German salami, Italian salami, Lebanon Bologna, pepperoni, Thuringer *cervelat*.
6. *Meat specialities.* Encompasses a wide variety of products that have in common only the fact that they are chopped or comminuted meats that are seasoned and usually cooked or baked rather than smoked. These are made from cured or non-cured meats and often made as loaves but usually sliced and served cold. Examples are chopped ham loaf, condiment loaf, head cheese, jellied corn beef, luncheon meat, minced ham, peppered loaf, scrapple and souse.

Kinsman finally mentions that, in addition, some of the products may be fermented, such as Lebanon Bologna and Thuringer, but that their basic formulation and processing is as described under category 5 (summer, dry, or semi-dry sausage). The two classification systems thus differ to some extent from one another.

Kinsman's book (1980) comprises a collection of sausage names and descriptions from meat scientists from 27 countries, with a total of 844 entries, from which 177 are classified as 'fermented'. Although the contributors were asked to use the same criteria in their classification system, a close examination of the entries reveals that it is only possible to state that nearly all countries produce fermented sausages, but they seem most popular in central European countries. In many countries, fermented sausages have names of very distinct German or Italian origin. Similarly the names of raw (country-style) hams which according to Leistner (1986b) are matured, for which reason they should be considered as belonging to the same category, very often are named according to the regions where they are produced. Names such as Westfalian, Ardennes, Parma or Spanish are thus well-known.

In his review, Adams (1986) explains that in the USA, fermented sausages are divided into two categories: dry and semi-dry or 'new condition' sausages. Originally these were defined in terms of the weight loss that occurred during processing or the final water content of the product. However, since the initial water content of the sausage preparation will vary according to the type and quantity of meat tissues used, an alternative system of classification based on the product's water : protein ratio was introduced. This correlated better with systems based on moisture content and the descriptive terminology frequently used by industry.

According to this terminology, fermented sausages may be classified as in the scheme shown in Table 3.2.

Table 3.2 Types of fermented sausage (modified from Adams, 1986)

Type	Weight loss (%)		Water content (%)		Water: protein ratio	Examples
Dry	25–50	30 (20 if smoked)	35	25–45	2.3:1	Pepperoni Milano and Genoa salami, French saucisson sec
Semi-dry	Medium dry, 30 New sausage, 20	20	50	40–50	2.3–3.7:1	Summer sausage, cervelat, Thuringer, chorizos
Undried (spreadable)		10				Teewurst Braunschweig, Mettwurst

Table 3.3 Classification of sliceable fermented sausages (from Roca and Incze, 1990)

Product and type	Production time	Water content (%)	Final a_w value*	Examples
spreadable	3–5 days	34–42	0.95–0.96	Teewurst, Frische, Mettwurst
sliceable				
Short ripening	1–4 weeks	30–40	0.92–0.94	Summer sausage
Long ripening	12–14 weeks	20–30	0.85–0.86	Hungarian salami, Italian salami, French saucisson sec

a_w = water activity.

3.3 Production methods and materials for fermented sausages

A third type of classification of fermented sausages was proposed by Roca and Incze (1990) as shown in Table 3.3. They also consider the time for fermentation/ripening.

Industrialisation has resulted in a number of changes in food production methods, including those used in the meat industry. If we look at production methods as used in the middle of the nineteenth century, we see that these were characterised by small production units lacking the necessary equipment and refrigeration. In addition, whereas today the raw materials for sausages and other meat products are typically taken from less valuable cuts (i.e. trimmings etc.), the meat raw materials were often taken from the more valuable cuts. A Danish professional book with recipes (Anon, 1861) thus indicates the following ingredients: Equal parts of beef from top round, lean pork meat, and back fat or similar fatty tissue; salt, pepper, cloves and allspice. The production method used meats which were finely hand-cut by knife and mixed for about 30 minutes. Then the meat was mixed with the salt and spices. The back fat was diced very carefully. Red wine or madeira could be added if the texture was too hard.

The mix was finally stuffed in casings with care being taken to avoid the formation of holes. The sausages were finally rubbed well with salt. After 24 hours, the mix was pressed together in the casing once more, and left for a further 24 hours, after which the salt was rubbed off and the sausages hung for smoking (cold smoke) for two to three weeks. In another recipe, the meat was scraped and/or sliced into thin slices. This recipe actually advises the adding of saltpetre and some brown sugar. Further, according to this recipe, the use of some beer was also common.

In another famous Danish professional book with recipes (Mangor, 1891), the recipes and production methods are almost identical to the above description. It is clear that under such circumstances, it is not possible to classify the sausages directly as 'fermented', although on some occasions they must have been subjected to vigorous bacterial growth during the drying period. Furthermore, although the salt levels of many north European 'fermented' sausages were higher than what they are today, they nearly always added sugar. Older instruction books (e.g. Schneekloth, 1913) stress on several occasions the need to work as hygienically as possible, 'in order to avoid the product to become rotten'.

Incze (1986) has given an account of the traditional production of Hungarian salami. As a curiosity, he mentions that the production of Hungarian salami became such a success during the nineteenth century because the demand for pork fat was so high that lean pork meat became very cheap. Up to about 30 years ago a rocker-knife was used for cutting meat. Today this has been replaced by better, automatic equipment such as the cutter produced by Krämer & Grebe. Proper cutting is of great importance in order to avoid smears on the cut meat surfaces. After cutting and mixing, the mix was left in vats for a few days before it was stuffed in natural casings. Today this step is completely avoided, since stuffing can be made after vacuuming the mix. It is essential to exclude any holes during stuffing, partly in order to improve colour stability and partly to reduce the redox potential in the mix to avoid any growth of aerobic microorganisms. Today the mix is usually frozen, or at least semi-frozen prior to stuffing, in order to avoid smears. Grinding can be made in a cutter, and in many cases the natural casing has been replaced by an artificial casing. Further drying, smoking and maturing takes place in climate chambers, where the temperatures and humidities can be adjusted as required. With a production time of 14–16 weeks, a final salt content of about 4.5%, and a_w water activity of about 0.93, it is essential to mature at a fairly low temperature. Incze (1986) reports a temperature of <10°C during most of the maturation and smoking time. From the information on the composition of the sausage it can easily be appreciated that before environmental chambers were available, such production could only be made during the winter.

Incze (1986) also remarked that, although a true Hungarian salami must be smoked, the original Italian fermented sausage was never smoked. The reason for this is probably the different climates in Italy and Hungary, i.e. during winter there are often humid periods which could support growth of moulds. Although it is now known how growth of moulds can be avoided, it was realised that the mouldy cover had a number of advantages including antioxidant properties and aroma formation. It is probably for this reason, that mouldy surfaces of the Hungarian sausage have been maintained.

During the last few decades much work on the use of curing salts in meat products has been carried out due to the concern regarding the toxic aspects of nitrite and nitrate. It is beyond the scope of this chapter to give a complete survey of this topic, but a few studies will be mentioned. Puolanne (1977) studied the effect of reduced addition of nitrite and nitrate on the properties of dry sausage. He concluded that growth of added lactic acid bacteria was retarded, if the concentration of added nitrite was 100 ppm and nitrate was 125 ppm. However, if the concentrations of nitrate and nitrite were halved, this promoted growth of lactic acid bacteria and resulted in a great pH drop. He also concluded that growth of micrococci was not affected by the concentrations of curing salts used in his studies. He further found that it was possible to obtain a normal colour of nitro-somyoglobin by addition of as little as 25 ppm nitrite and 50 ppm nitrate.

Another finding in the study was that if the fermentation temperature was reduced from 26 to 18°C, the reduction time of nitrate to nitrite was slowed-down, and the end pH was increased. He concluded that if a mixed bacterial starter culture of *Lactobacillus plantarum* and *Micrococcus* spp. and a nitrate concentration of 150 ppm, or 50 ppm nitrite plus 75 ppm nitrate was used, he was unable to distinguish sensory differences regarding aroma and taste, or colour, on the cut surface of the sausages. In a study on fermented American sausage, Dethmers *et al.* (1975) also concluded that addition of 50 ppm nitrite was necessary to achieve typical flavour and appearance. Puolanne (1977) was also unable to find any effect of added nitrite/nitrate concentrations on growth of *Salmonella senften-berg*. This is in contrast to Schmidt (1990), who found growth of *Salmonella* in Mettwurst, if the salt content was 2.5%, the nitrite content 60 ppm, and the maturation temperature 25°C. Tjaberg *et al.* (1974) concluded from a study on the production of fermented sausages with or without nitrite, that there was no difference in the microflora in fermented sausages. This could be due to the presence of nitrite, but at a storage temperature of 12°C a sensory evaluation showed reduced shelf life, if the sausages were produced without nitrite. The colour was not assessed during the study.

3.4 Microbial starter cultures

The intrinsic factors of a sausage mix possess a natural selectivity for promoting the development of the desired microflora. A traditional means of ensuring that the proper microbial flora was present was to add up to 5% of a previous mix to a fresh batch, the so-called 'back-slopping' method, which of course has been known for centuries from, for example, bread manufacturing. Correctly applied, this technique works well and allows an element of strain selection, since good batches may be retained as multiple inocula. Bacus and Brown (1981) reported that the back-slopping technique, as well as the traditional process of relying on the natural controls to ensure a natural fermentation by the indigenous bacteria are still commonly practised today. However, due to the increasing use of starter cultures which had taken place with such great effect in the dairy industries, attempts were made to develop starter cultures for other foods, such as meats. Jensen and Paddock (1940) were probably the first to have published and patented the idea, using a *Lactobacillus*. In his book, Jensen (1945) reports how, through bacteriological examination of special types of acid 'tangy' sausage, such as Thuringer sausage, he disclosed the fact that at least six *Lactobacillus* species could be isolated singly or in combinations. Once isolated, mother cultures were made, and these cultures could be inoculated in new batches of sausage trimmings. Through this method, it was possible for sausage manufacturers to gain tremendous savings because less-effective chance inoculations were eliminated. He continues by reporting on a Bologna sausage with a very characteristic flavour and which could not be prepared in any place except in a certain Pennsylvanian plant. The *Lactobacillus* responsible for the characteristic flavour, colour, and texture was finally isolated so that similar Bologna sausages could be successfully manufactured in any plant. The use of a defined microbial culture thus originated in the USA with a series of patents issued during the period 1920–40.

According to Bacus and Brown (1981), subsequent attempts to use *Lactobacillus* were not very successful. Although these bacteria were dominant in fermented sausages and indeed the predominant strains on the market, *Lactobacillus* was not very well suited for use at that time because it did not readily survive lyophilisation which was the chief method for distributing starters. For this reason, Deibel *et al.* (1961), cited by Bacus and Brown (1981) introduced *Pediococcus cerevisiae* as the commercially available meat starter culture. Although it was not encountered as frequently in the naturally fermented products, it appeared to meet the requirements of a sausage starter culture. When better methods for freeze-drying of *Lactobacillus* strains were developed and with the advent of frozen culture concentrates, the use of *Lactobacillus* became common again.

The European era of using bacterial starter cultures for meat products is undoubtedly due to the research carried out by Finnish workers, first of all Niinivaara. In his thesis, Niinivaara (1955) explains some of the reasons why his work was initiated: contrary to the reasons by American workers, who first and foremost wished to lower the pH and ensure the 'tangy' taste and shelf life, Niinivaara emphasised the importance of proper colour formation. Only as a secondary point did he mention the formation of hydroxylamine as a preservative. He therefore made sure of isolating bacterial cultures which were good nitrate reducers, and therefore, isolated primarily *Micrococcus* cultures from Finnish curing brines and selected a strain which fitted well with the proposed selection criteria. The criteria were: (i) growth in 15% sodium chloride; (ii) optimum temperature for growth 25–30°C; (iii) non-pathogenic; (iv) pH optimum 6.6–7.0; (v) reduces nitrate to nitrite. The selected strain was later known as the *Micrococcus* M53 strain which also produces hydroxylamine and lactic acid. Later researchers have emphasised the importance of micrococci in fermented sausage for their ability to reduce nitrate to ensure colour formation, and their catalase activity which destroys peroxide (and which can cause fading of the curing colour from red to grey or brown, or even green discoloration of nitrosomyoglobin at high concentrations of peroxide in the absence of catalase) (Niven and Evans, 1957). Others, such as Demeyer *et al.* (1974) have stressed that the micrococci are the main agents of lipid degradation and formation of carbonyls, thereby contributing to the aroma of fermented meats.

After the publication of Niinivaara's thesis in 1955, a number of Finnish publications appeared on the subject, to a large extent stressing the usefulness of *Micrococcus* in fermented sausages. These publications include those by Niinivaara and Pohja (1956, 1957) and the thesis by Pohja (1960). In his introduction, Pohja stresses that micrococci are superior to lactobacilli in their versatility as producers of desirable characteristics through their contribution to the improvement of colour, aroma, flavour and keeping quality of dry sausages. They also constitute the most important group of bacteria in regard to their activity in the processing of dried meat products.

In a later Finnish publication (Nurmi, 1966), the results of various inoculation experiments were reported. Nurmi found that when he inoculated with lactobacilli alone, there was a rapid pH decrease and a rapid development of the desired consistency, presumably as a consequence of the pH drop. However, he also observed serious discoloration and flavour defects in several experimental series, probably due to accumulated hydrogen peroxide formed by the lactobacilli. Finally, he reported high peroxide numbers in the sausages. When using micrococci, he only found a moderate pH drop and a slight colour formation compared to a non-inoculated control. Best results were obtained, when he used a mixture of *Micrococcus* and *Lactobacillus* spp. as a starter culture. In his experiments he used nitrate in most cases and reported, that when nitrate was substituted with nitrite, he could prevent discolorations, but not flavour faults. This, he noted, could be avoided, if micrococci were inoculated in the sausages. Nurmi also tried inoculation with *Pediococcus cerevisiae*, and found some acceleration in the pH decrease. However, he also found colour and flavour defects in the finished sausages in some of his experiments.

The idea of using mixtures of starter cultures in fermented meats was subsequently commonly used in practice later. Today, several companies offer various blends, primarily comprising mixtures of micrococci and various kinds of lactic acid bacteria. Perhaps the most well-known commercial type of mixture is Duplofermente 66, which comprises *Lactobacillus plantarum* and *Micrococcus* sp. (Rudolf Muller, Giessen, Germany). However, other mixtures are also marketed, which are blends of various mixes of lactic acid bacteria to enable the sausage manufacturers to use a broad range of fermentation temperatures (Bacus and Brown, 1985).

Over the years several attempts have been made to use species other than micrococci and various lactic acid bacteria as starter cultures. Thus it has been a very old tradition in many countries to 'inoculate' new curing brines with samples of old brines, in particular if meat products cured in new brines were of poor quality. One example dates back to Denmark in the 1930s, where a bacon factory was burned down. Because it was virtually impossible to produce proper bacon in the new bacon factory, which was built after the fire, some barrels of old curing brine from a neighbouring bacon factory was used to inoculate the new brine. This solved the problem completely (Madsen, 1963). Leistner (1960) (cited by Pätäja, 1977) examined the effect of bacteria in brines made by inoculation of old brines and reported that although micrococci and lactobacilli do improve the stability of curing brines and prevent the spoilage of brines, bacteria belonging to the genera *Vibrio*, *Spirillum*, *Acromobacter*, *Alcaligenes* and *Micrococcus* could improve colour formation, and that mixtures of these bacteria had a beneficial effect on the flavour. Others, e.g. Buttiaux (1957), isolated *Vibrio costicolus* which was used in meat curing in France, primarily as a nitrate reducer, but also because the strain was considered as important for aroma and flavour formation. These were properties which were also emphasised by Hawthorn and Leitch (1962). Similarly, Pätäja (1972) (cited by Pätäja, 1977) reported experiments in which he had found *Vibrio costicolus* and *Acromobacter* strains which were useful in meat curing, both in cover brine and especially in the brine for injection into meat. The strains were used as starter cultures in dried hams and improved both colour, texture and flavour of the finished product.

In his thesis, Pätäja (1977) concludes that in all his experiments, samples prepared without starter cultures were, as a rule, of the poorest quality. In some cases he succeeded in the production of dry sausages of a high quality when he used a specific *Aeromonas*, but more often than not, when using other strains of Gram-negative bacteria alone, such as his isolate *Vibrio* 21, *Escherichia coli* and *Proteus vulgaris*, no beneficial effect could be observed. However, to be of satisfactory quality, Finnish fermented (dried) sausages must have a low pH, a property which would be difficult to obtain with most Gram-negative starter cultures.

Over the years, several new attempts have been made to use Gram-negative bacteria as starter cultures or to investigate any beneficial effect these bacteria might have on the flavour of meat products, such as bacon. Gardner (1980) thus mentions this. In a review (Lücke and Hechelmann, 1987), they describe much of the work done over the years with Gram-negative bacteria as aroma enhancers in fermented meat products. They conclude, however, that in their opinion Gram-negative bacteria will probably never be commercially available as starter cultures, since they are very difficult to freeze-dry, store and distribute, and if they are incorrectly handled, they may cause spoilage in the products instead of aroma improvements. Attempts to use *Vibrio* as aroma improvers are still being made (Andersen and Hinrichsen, 1991).

3.5 Yeast starter cultures

Although lactic acid bacteria and micrococci are the predominant microorganisms used for fermented meats, yeasts have also been tried, since they often have been isolated from cured and fermented meats. Thus, yeasts were commonly found on stored dry salami in the past, when these products were distributed unwrapped or vacuum-packed.

Some starter culture preparations contain yeasts. Early workers such as Rossmanith *et al.* (1972) (cited by Lücke and Hechelmann, 1987) reported that *Debaryomyces hansenii*, identified as *Debaryomvces kloeckeri* was found to give the best performance in dry sausage ripening. In his review on the microbiology of fermented sausages, Coretti (1973) reports that rapid and stable development of red colour and acceptable aroma could be obtained with *Debaryomyces kloeckeri*, *D. cantarelli* or *D. pfaffii* as well as with a mixture of micrococci, lactic acid bacteria and *D. hansenii*. Yeasts have also been used as starter cultures by some eastern European researchers. Miteva *et al.* (1986) reported on the use of *Candida utilis*. Similarly, Gehlen *et al.* (1991) have reported on the influence of *Debaryomyces hansenii* together with lactic acid bacteria and micrococci. In two reports on *soudjouk*, a fermented Turkish sausage made from beef and mutton, Goekalp (1985, 1986), also stresses how the use of various mixtures of starter cultures can be improved by adding *D. hansenii*.

3.6 Mould starter cultures

As mentioned earlier, moulds are also employed as starter cultures. Moulds on fermented meats were originally probably tolerated rather than desired. Incze (1986) mentions that the Hungarian salami, which probably is the best-known fermented sausage with moulds, was originally made without moulds, when it was manufactured in Italy, but due to the climatic conditions, moulds became a characteristic feature of these sausages. Originally the so-called 'house flora' just contaminated the products, but when the significance of mycotoxins was realised, much development work was carried out in order to avoid the use of toxinogenic moulds. Much of this work was carried out at the Federal Centre for Meat Research in Kulmbach, Germany. Work by Leistner and Echardt (1979) has also shown that many mould isolates from Hungarian sausages were toxinogenic *Penicillium* strains. Mintzlaff and Leistner (1972) had already isolated some mould starter cultures, which were shown to be non-toxic. These included a strain of *Penicillium nalgiovense*, which was later known as 'Edelschimmel Kulmbach', Leistner *et al.* (1980) further isolated *Penicillium chrysogenum* which showed a green mycelium at 20°C, but not at 10°C. According to Lücke and Hechelmann (1987), a French strain of *P. chrysogenum* is being marketed, a mutant, which is white under all growth conditions on fermented sausages.

3.7 Raw hams

It can be debated if raw hams belong to the category of fermented meats. Certainly, if the term 'fermentation' is restricted to microbial breakdown of the ingredients in meat, hams would not qualify. Thus, many studies have shown that the numbers of microorganisms are low compared with fermented sausages. Giolitti *et al.* (1971) showed that the total plate count was below 10^5/g of lean meat during a storage and 'maturation' period of six months, and the numbers of halophilic and halotolerant bacteria did not exceed 2.5×10^6/g. Furthermore, although a fermented sausage will be subjected to temperatures above 10–15°C and often above 20°C during manufacture, a true 'raw ham', 'country style ham' or hams with similar names will be subjected to much lower temperatures. Leistner (1986b) thus cites Baldini (personal communication) who recommends that Parma hams must be stored at temperatures below 4°C during the first several months. Higher temperatures — up to 20°C — are not used until after six to seven months. However, in the same article, Leistner stresses that improvements in the production of raw hams will most likely involve bacterial starter cultures, but he warns against relying on this. In his opinion, time and temperatures during production remain the decisive factors in a proper ham manufacture. Another peculiarity in raw ham manufacture is that nitrite and nitrate are omitted in many cases.

One way to shorten the production time of raw hams and yet promote aroma formation is to smoke the products. Pending regional habits, the smoking temperatures may vary from 15 to 25°C, or even up to 30°C. Smoking is considered an important process in order to avoid mould growth, and is therefore especially employed in areas with comparatively high humidities.

As mentioned earlier, the art of producing fermented sausages is old. The art of producing raw hams, and other cuts of preserved raw meats, is much older. According to Leistner (1986b), who quotes Kroll and Witte (1921) raw hams are mentioned by both Cato in 160 BC, as well as by the Roman Emperor Diocletian in AD 301. Similarly, Varro, a Roman army commander and later governor, who lived in the first century, reports that the trade with, and manufacture of, raw hams was very popular throughout Europe in his time. Raw hams are also known in other cultures, such as China, as well as in many more primitive cultures, where drying was the principal mode of preservation. The original production methods probably always involved the manufacture of bone-in hams, although other cuts of fermented and dried whole meats have been, and are, known. The original procedures used for manufacturing of raw hams have always been to dry-cure the meat and hang it at low temperatures. More recently, light brine pumping has been used in some cases. The variations in production methods have mostly consisted of variations in the addition of different spices.

It is generally accepted that the changes in flavour which take place in raw hams are primarily due to breakdown by endogenous enzymes. These processes are not yet fully understood yet, although several research groups are studying aroma formation in raw hams (e.g. Giolotti et al., 1971; Buscailhon et al., 1991; Toldra et al., 1991).

3.8 Closing remarks

In addition to the preserving effect which fermentation has on a number of commodities, a fact man has made use of for thousands of years, the aroma of fermented foods is also regarded as important. Further, fermented foods have been enjoyed for many years because of the health aspects usually attributed to them. At the beginning of this century, Metchnikoff thus maintained that the health-giving properties of fermented milk were 'magical' and could directly prolong life of man. Research during this century has revealed that there are several health aspects of fermented foods. Huis in't Veld et al. (1990) reviewed a number of health aspects relating to fermented foods. These include increased digestibility, therapeutic properties such as ability to reduce the level of serum cholesterol, an antitumour activity and effects on the immune system, as well as bacteriocins and probiotic activity formed in fermented foods. Many of these properties are not present in fermented meats, but a most important claim is that the fermenting microorganisms compete with pathogens. No doubt this may have prevented many cases of food poisoning, particularly in earlier times.

However, there are also several examples showing how ingestion of fermented foods can be dangerous. They include the risk of consuming mycotoxins and amines, which have been formed during the fermentation period. Nevertheless, there seems to be no doubt that in the consumer's mind, fermented foods possess an image of wholesomeness, an attribute which will be very useful for this class of food, which undoubtedly will have a great future, with many new products.

Over the years, the producers of fermented meats and other fermented foods have been hampered by the long production time involved. This has increased the price of the finished products, so naturally much development work has been aimed at reducing the time of manufacture. It is well-known that various acidulants are being used as additives, but so far it seems generally agreed by the producers and consumers alike that the production of high-quality fermented foods takes long time. Therefore, it seems that the time reduction during production is today only modest and is due to the fact that optimal temperatures and humidities can now be applied mechanically, but that fermentation and 'maturation' processes do take time. Leistner (1986b) warns against uncritical use of *Schnellmetoden* (quick methods). Historically, it therefore seems that much information is available on the production of fermented foods, but far from all compounds formed during fermentation have been identified. As far as new fermented meats and meat products are concerned, we are most likely only at the beginning of a new era of a class of foods and food ingredients.

References

Adams, M.R. (1986) Fermented flesh foods. *Progr. Indust. Microbiol.* **23**, 159–198.

Andersen, H.J. and Hinrichsen, L.L. (1991) Growth profiles of *Vibrio* species isolated from Danish curing brine. *Proc. 37th Int. Congr. Meat Science* Technol., Vol. 2, 4:2, pp. 528–533.

Anon (1861) *Koge og syltebog*, pp. 546–548. Th. Gandrup, Copenhagen.

Bacus, J.N. (1984) Historical perspective. *Utilization of Microorganisms in Meat Processing*, pp. 1–14. Research Studies Press, Letchworth.

Bacus, J.N. and Brown, W.L. (1981) Use of microbial cultures: meat products. *Food Technol.*, **35**, 74–78.

Bacus, J.N. and Brown, W.L. (1985) The lactobacilli: Meat products. In *Bacterial Starter Cultures for Foods*, pp. 57–72. CRC Press Fl.

Buscailhon, S., Touraille, C., Lacourt, A., Girard, J.P. and Monin, G. (1991) Relationships between tissue composition and sensory qualities of dry cured ham. *Proc. 37th Int. Congr. Meat Sci. Technol.*, Vol. 2, 6:5, pp. 859–862.

Buttiaux, R. (1957) Technique simple d'examens bacteriologique des saumures de jambon et ses résultats. [The microbiology of fish and meat curing brines.] *Proc. II Int. Symp. Food Microbiol*, pp. 137–146. HMSO, London.

Coretti, K. (1971) *Rohwurstreifung und Fehlerzeugnisse bei der Rohwurstherstellung*. Verlag der Rheinhessischen Druckwerkstatte E. Dietl W. Alzey.

Coretti, K. (1973) Warum interessiert den Praktiker die Mikrobiologie der Rohwurstreifung? *Fleischwirtschaft*; **53**, 907–908, 910–911.

Demeyer, D., Hoozee, J. and Mesdom, H. (1974) Specificity of lipolysis during dry sausage ripening. *J. Food Sci.*, **39**, 293–296.

Dethmers, A.E., Rock, H., Fazio, T. and Johnston, R.W. (1975) Effect of added sodium nitrite and sodium nitrate on sensory quality and nitrosamine formation in Thuringer sausage. *Food Sci.*, **40**, 491.

Gardner, G.A. (1980) Identification and ecology of salt-requiring *Vibrio* associated with cured meats. *Meat Sci.*, **5**, 71.

Gehlen, K.H., Meisel, C., Fischer, A. and Hammes, W.P. (1991) Influence of the yeast *Debaryomyces hansenii* on dry sausage fermentation. *Proc. 37th Congr. Meat Sci. Technol.*, Vol. 2, pp. 871–876.

Giolitti, G., Cantoni, C.A., Bianchi, M.A. and Renon, P. (1971) Microbiology and chemical changes in raw hams of Italian type. *J. Appl. Bact.*, **34**, 51–61.

Goekalp, H.Y. (1985) Turkish-style fermented sausage (*soudjouk*) manufactured by adding different starter cultures and using different ripening temperatures. I. Growth of total, psychrophilic, proteolytic and lipolytic microorganisms. *Fleischwirtschaft*, **65**, 1235–1240, 1248–1250, 1253–1254.

Goekalp, H.Y. (1986) Turkish-style fermented sausage (*soudjouk*) manufactured by adding different starter cultures and using different ripening temperatures. II. Ripening period, some chemical analysis, pH values, weight loss, colour values and organoleptic evaluations. *Fleischwirtschaft*, **66**, 573–575.

Hawthorn, J. and Leitch, J.M. (1962) *Recent Advances in Food Science*. Vol 2, pp. 281–282. Butterworths, London.

Huis in't Veld, J.H.J., Hose, H., Schaafsma, G.J., Silla, H. and Smith, J.E. (1990) Health aspects of food biotechnology. In *Processing and Quality of Foods*, Vol. 2. Eds P. Zeuthen, J.C. Cheftel, C. Eriksson, T.R. Gormley, P. Linko and K. Paulus, pp. 73–97. Elsevier Applied Science, London.

Incze, K. (1986) Technologie und Mikrobiologie der ungarischen Salami. *Fleischwirtschaft*, **66**, 1305–1311.

Jensen, L.B. (1945) Controlled inoculation. *Microbiology of Meats*, 2nd edn. The Garrard Press, Champaign, IL.

Jensen, L.B. and Paddock, L.S. (1940) US Patent 2,225,783.

Kinsman, D.M. (1980) *Principal Characteristics of Sausages of the World Listed by Country of Origin*. Copyright Professor Donald M. Kinsman, Animal Industries, University of Connecticut, Storrs, Ct.

Kramlich, W.E. (1971) In *The Science of Meat and Meat Products*. Eds J.F. Price and B.S. Schweigert, p. 484. W.H. Freeman, San Francisco, CA.

Kroll, W. and Witte, K. (1921) *PAULYS Realenzyclopädie der classischen Altertumswissenschaft*. Alfred Druckenmüller, Stuttgart.

Leistner, L. (1960) Microbiology of ham curing. *Proc. 12th Conf. American Meat Institute Foundation*, Circular, 61, 17–23.

Leistner, L. (1986a) Allgemeines über Rohwurst. *Fleischwirtschaft*, **66**, 290–300.

Leistner, L. (1986b) Allgemeines über Rohschinken. *Fleischwirtschaft*, **66**, 496–510.

Leistner, L. and Echardt, C. (1979) Vorkommen toxinogener Penicillien bei Fleischerzeugnissen. *Fleischwirtschaft*, **59**, 1892–1896.

Leistner, L. Hechelmann, H. and Trapper, D. (1980) *Penicillium chrysogenum*, eine Starterkultur für Rohwurst. *Jahresbericht. Bundesanstalt für Fleischforschung, Kulmbach*, C22.

Lissner, E (1939) *Wurstologia oder es geht um die Wurst*. Hauserpresse, Hans Schaefer, Frankfurt am Main.

Lücke, F.K. (1985) Fermented sausages. In *Microbiology of Fermented Foods*, Vol 2. Ed. B.J.B. Wood, pp. 41–83. Elsevier Applied Science, London.

Lücke, F.K. and Hechelmann, H. (1987) Starter cultures for dry sausages and raw ham. Composition and effect. *Fleischwirtschaft*, **67**, 307–314.

Madsen, J. (1963) Personal communication.

Mangor, A.M. (1891) *Kogebog for smaa Husholdninger*, pp. 148–150. Th. Lind, Copenhagen.

Mintzlaff, H.J. and Leistner, L. (1972) Üntersuchungen zur Selektion eines technologisch geeigneten und toxikologisch unbedenklichen Schimmelpilz-Stammes für die Rohwurst-Herstellung. *Zeitbl. Vet. Med. B*, **19**, 291–300.

Miteva, E., Kirova, E, Gadeva, D. and Radeva, M. (1986) Sensory aroma and taste profiles of raw-dried sausages manufactured with a lipolytically active yeast culture. *Nahrung*, **30**, 829–832.

Niinivaara, F.P. (1955) Über den Einfluss von Bakterienreinkulturen auf die Reifung und Umrotung der Rohwurst. *Acta Agralia Fennica*, **85**, 1–128.

Niinivaara, F.P. and Pohja, M.S. (1956) Über die Reifung der Rohwürst. I Mitteilung. Die Veränderungen der Bakterienflora während der Reifung. *Zeitschr. Lebensmittel-Untersuch. Forsch.*, **104**, 413–422.

Niinivaara, F.P. and Pohja, M.S. (1957) Erfahrungen über die Herstellung von Rohwurst mittels einer Bakterienreinkultur. *Fleischwirtschaft*, **9**, 264–286.

Niven, C.F. Jr. and Evans, J.B. (1957) *Lactobacillus viridescens* nov. spec., a heterofermentative species that produces a green discoloration of cured meat pigments. *J. Bact.*, **73**, 758–59.

Nurmi, E. (1966) Effect of bacterial inoculations on characteristics and microbial flora of dry sausage. *Acta Agralia Fennica*, **108**, 1–77.

Pätäja, E. (1972) Starterkulturen bei der Pökelung. *Berichtung Starterkultur Symposium*, pp. 169–179. Institute for Meat Technology, University of Helsinki.

Pätäja, E. (1977) The effect of some Gram-negative bacteria on the ripening and quality of dry sausage. *J. Sci. Agric. Soc. Finland*, **49**, 107–166.

Pohja, M.S. (1960) Micrococci in fermented meat products. *Acta Agralia Fennica*, **96**, 1–80.

Puolanne, E. (1977) Der Einfluss von verringerten Nitrit und Nitratzusatzen auf die Eigenschaften der Rohwürst. *J. Sci. Agric. Soc. Finland*, **49**, 1–106.

Roca, M. and Incze, K. (1990) Fermented sausages. *Food Rev. Int.* **6**, 91–118.

Rossmanith, E., Mintzlaff, H.J., Streng, B., Christ, F. and Leistner, L. (1972) Starterkulturen für Rohwürste. *Jahresbericht der BAFF*, 147–148.

Schmidt, U. (1990) Kontrolle von Lebensmittelvergiftern bei der Herstellung von Fleischerzeugnissen. Sichere Produkte bei Fleisch und Fleischerzeugnissen. *Bundesanstalt Fleischforsch. Kulmbacher Reihe*, Band 10, 44–68.

Schneekloth, H.F. (1913) *Økonomisk Vejleder for Kød-Industrien*. Published by the author in Copenhagen.

Smith, D.R. (1987) Sausage — a food of myth, mystery and marvel. *CSIRO Food Res. Quart.*, **47**, 1–8.

Tjaberg, T.B., Skjelkvåle, R. and Valland M. (1974) Produksjon av spekepløse med og uten nitritt. *NINF Informasion*, **3**, 1–39.

Toldra, F., Motilva, M.J., Rico, E. and Flores, J. (1991) Enzyme activities in the processing of dry-cured ham. *Proc. 37th Int. Congr. Meat Sci. Technol.*, vol. 3, 6:28, pp. 954–957.

4 Bacterial fermentation of meats

L. KRÖCKEL

4.1 Introduction

As with all dead organic matter, muscles from slaughtered animals, as a whole or in particulate form, may be modified by microorganisms during prolonged storage. Environmental conditions and storage time greatly influence the sort and extent of modification. For foods, there can be desirable and undesirable microorganisms, which bring about desirable and less-desirable changes. Desirable modifications are improvements in flavour, aroma, palatability, appearance and storage characteristics. Microbial activities that result in off-odours, strange taste, health-threatening metabolites, colour deterioriation, loss of consistency, and the growth of pathogenic and toxinogenic bacteria generally spoil the food and thus make it unsuitable for human consumption. Changes in microbial abundance, water activity (a_w), pH, pO_2 and concentration of chemical compounds may be due to real fermentation processes, i.e. incomplete anaerobic oxidations of organic substrates, or to aerobic microbial metabolism. However, it is common to call a food 'fermented' if microorganisms, be they truly fermenting or not, or enzymes had contributed significantly to its final characteristics (Campbell-Platt, 1987). Several meat enzymes are known to remain active in dead muscles, e.g. the glycolytic sequence, lipases and proteases. Dead or non-growing microorganisms may release or provide active enzymes, e.g. nitrate reductase and catalase. Meats may also be tenderised by the action of certain plant enzymes. However, although the *post mortem* pH drop in meat is the result of a true indigenous fermentation by meat enzymes, this meat is not generally considered fermented. In contrast, dry-cured raw hams which are mainly modified by the action of lipolytic and proteolytic meat enzymes are often considered 'fermented meats'. Similarly, 'fermentation' and 'ripening' are often used synonymously, although the last expression may appear to reflect more an 'aging' process without participation of microorganisms or after microbial activities have slowed down. Finally, a food in which undesired fermentation products have accumulated is not usually considered fermented but spoiled. Whether a fermentation is desirable or not also depends on the acceptance by the consumer. Mould-ripened salami varieties are appreciated as a delicacy especially by Mediterranean consumer groups, while some northern European and US consumers do not readily accept the appearance and smell of these products. Organoleptic compounds of microbial origin may be desired or not, depending largely on their final concentrations in the specific products. Thus, spoilage and desired fermentation of foods by microorganisms are very closely related.

4.2 Fermented foods

A wide variety of organic materials are subjected to a fermentation or ripening process involving the participation of different microorganisms to produce acceptable and stable foods. The majority of these are fermented by lactic acid bacteria (LAB) although other microorganisms, e.g. yeasts, may be involved. Species and strains of the genera *Lactobacillus, Leuconostoc, Pediococcus* and *Streptococcus* are found or used in fermented foods. The raw materials for fermented foods include various meats, fish, milk, cereals, vegetables, soya, peanuts, cocoa beans and coffee berries. Many fermented foods have developed from natural fermentations, with the traditional techniques selecting a particular lactic flora by means of salting, temperature control or 'back slopping', i.e. inoculation from a previous batch. This has been developed commercially into the selection and use of highly specific starter strains, particularly for dairy products but also for meat products and currently for fish silages and sour dough breads among others. In general, the LAB are selected on the basis of rapid acid production under a wide range of conditions and/or batches of raw materials (Hesseltine, 1984; Gibbs, 1987; Raccach, 1987). The use of LAB in fermented foods has been the subject of various reviews (Bacus and Brown, 1981; Liepe, 1983; Bacus, 1984; Gilliland, 1985; Lücke, 1985; Wood, 1985; Adams, 1986; Leistner, 1986; Leistner and Lücke, 1989).

4.3 Fermented meats

Most types of raw sausages and raw hams are modified by microorganisms and are thus 'fermented' (Leistner, 1991). Today a large variety of fermented sausages are being produced. They include fast, moderately and slow ripened products, the ripening time being mainly a function of temperature, starter culture and carbon source. The greatest variety of raw sausages is found in Germany. It ranges from short to long ripening times, from coarse to fine meat-particle sizes and from small to large diameters. There are also differences in acidification (final pH 4.7–5.5), flavour and firmness. In the north, generally more sour sausages are produced than in the south. Spreadable varieties are common. In Austria, natural casings and lower amounts of sugar are used. The sausages are less sour (pH may be above 6.0, for *salametti* even 7.4) and more intense in aroma. In France, mean-to-coarse comminuted meat is ripened with small amounts of sugar for a long time in natural casings and flavoured with fungi and yeasts. The pH values are generally above pH 6.0. Switzerland has a similar broad variety of sausages as Germany but is oriented more towards French-Austrian- and Italian-style sausages with pH values above 5.5 (Frey, 1987). For mould-ripened sausages of the Hungarian or Bulgarian type *Penicillium expansum* and *P. viridicatum* are used, while in France, Germany, Spain and The Netherlands, *P. nalgiovense* is preferred. Ingredients may vary widely, due to specific country regulations. For example, *chorizo* sausages show a wide variation in fat and moisture content. Many samples contain added phosphates, since Spanish law permits 5000 ppm P_2O_5 in the raw material and 8000 ppm in the finished product (Botas *et al.*, 1987). The 'typical' summer sausage recipe contains no nitrate or nitrite but very high amounts of sodium chloride (6%) and glucose (4.5%) as compared to European sausages (Burrowes *et al.*, 1986). *Nham* is a quite different fermented meat product, made from raw pork and traditionally wrapped in small banana leaf packets. It is very popular in Thailand and common to South-East Asia (Petchsing and Woodburn, 1990). Chinese sausages, like *lup cheong* belong to the class of non-fermented dry sausages which are mainly preserved by drying (Leistner, 1988; Liepe, 1988; Savic *et al.*, 1988).

Raw dried hams are the result of a long enzymatic maturation in the presence of salt, essentially by the meat enzymes themselves. Classical Italian ham (*prosciutto di Parma*) has a ripening time of 10 months. The activities of indigenous lipases and proteases contribute significantly to final flavour. The role of bacteria is less clear, since curing temperatures around 2–6°C and very high salt levels are quite unfavourable for microbial activities. In nitrate-cured hams Micrococcaceae usually dominate the flora of the curing brine. Reduction of nitrate to nitrite is probably the most significant bacterial contribution in the ripening process (Poma, 1987; Jacquet, 1988; Leistner, 1991).

4.4 Meat as a substrate

During slaughter, meat often becomes contaminated with all kinds of microorganisms present in the immediate environment. These include both fermenting and non-fermenting bacteria. Due to the highly nutritive nature of meat, a vast variety of organotrophic bacteria will be able to grow. A water activity (a_w) close to 0.99 and a pH around 7.0 as found in fresh meat coincides with the optimal growth conditions of most bacteria. Storage conditions basically determine what kinds of microorganisms will develop. Therefore effective measures must be taken to prevent spoilage or growth of toxinogenic and pathogenic microorganisms. If, after slaughter, fresh meat is stored immediately at 0–2°C microbial growth will be almost negligible during the first 24 hours. The pH usually drops down to 5.7–6.0, due to the fermentative metabolism of glycogen to lactic acid by the indigenous meat enzymes. However, this pH is still high enough to allow the development of the neutrophilic 'cold room flora', consisting mainly of aerobic pseudomonads and facultative anaerobic enterobacteria. Their proteolytic activities will again increase the pH on the meat surface, as far as unpackaged meat is concerned. Under the microaerophilic conditions found in packaged meats these proteolytic, acid-sensitive microorganisms are competed out by the psychrotrophic lactobacilli and *Brochothrix thermosphacta* which lower the pH by acidic fermentation of residual carbohydrates. Therefore longer storage times are possible with vacuum-packaged meat. It is usually just high-quality beef that, for the purpose of tenderising, is stored aerobically or vacuum packaged over a prolonged period. The process of tenderising is mainly due to indigenous meat enzymes, but metabolic end-products of microorganisms may contribute to the final organoleptic properties of ripened meats.

Other kinds of meat are generally processed within one week after slaughter and are sold directly as fresh meat to the consumer, or they are manufactured — with or without curing — into a variety of cooked and uncooked meat products such as sausages and hams. Cooking aims to inactivate all spoilage microorganisms and pathogens with a minimum destruction of product integrity, taste and nutritive value. However, cooked products can become recontaminated with microorganisms during post-cooking treatment. Due to the absence of a competitive flora and a better access to nutrients, the composition of the bacterial flora on these processed foods is completely different from that on raw meats.

Uncooked meat products are usually subjected to curing with curing salts, containing nitrite or nitrate. Besides physico-chemical effects on the meat constituents, curing establishes selective conditions for the microorganisms by lowering the a_w and, if nitrite is used, impeding the nitrite-sensitive microorganisms like salmonellae. However, recent evidence suggests that the amounts of nitrite used in fermented sausages do not inhibit the growth of *Salmonella* spp., *Clostridium botulinum* and *Staphylococcus aureus* in these environments (Lücke, 1985). The small amounts of nitrite (120–150 ppm) used for meat curing may only be effective when combined with other growth impeding factors such as reduced a_w and pH. In cured sausage meats (a_w 0.96–0.97) the Gram-negative flora is mainly affected by the lowered a_w. These water activities greatly favour the development of *Micrococcaceae* and LAB. Nitrate, unless it is reduced to nitrite, mostly by members of the genus *Micrococcus* and *Staphylococcus*, is not toxic to microorganisms. High salt levels, used during curing of air-dried and smoked raw hams, select even more Micrococcaceae and other halotolerant microorganisms such as *Vibrio* and *Halomonas*. LAB never reach high numbers in these environments. Consequently, the final pH of ripened raw hams is usually identical to the starting pH.

4.5 Bacteria

Within the eubacteria, fermentative metabolism is found in the entero-bacterial and the clostridial tribe. All LAB of relevance in food technology are members of the phylogenetically defined division of Gram-positive organisms. Essentially two bacterial groups are responsible for the desired meat fermentations that lead to safe and stable products, the LAB and the Micrococcaceae. All other bacteria found in meats are usually related to undesired fermentations and/or spoilage, e.g. Enterobacteriaceae, *Brocho-thrix thermosphacta* and *Pseudomonas*, whereas certain species of *Salmon-ella, Listeria* and *Clostridium* are highly undesirable because of their potential pathogenicity. In meat fermentations, Gram-negative bacteria are usually present only in small numbers which rapidly decline over time. However variations in meat quality may introduce different amounts of normally undesirable bacterial degradation products that contribute to a specific, eventually unacceptable, flavour.

4.5.1 Lactic acid bacteria (LAB)

The LAB commonly found on meat belong to the genera *Lactobacillus, Pediococcus, Streptococcus* and *Leuconostoc*. Lactobacilli usually domin-ate the LAB flora in naturally fermented sausages and vacuum-packaged meats. Besides, they are mainly found in the intestinal tract of healthy humans and animals as well as in fermenting vegetables or plant materials, such as silage. Rod-shaped LAB are largely used in the preparation of a variety of foods and feed products (Bottazi, 1988). *Lactobacillus plantarum* is normally associated with meat and vegetable fermentations, and is frequently isolated from white pickled cheese (El-Gendy *et al.*, 1983). *Lactobacillus plantarum, L. brevis, L. farciminis, L. alimentarius* and *L. curvatus* are essential for the fermentation of sausages prepared with added sucrose or powdered milk (Reuter, 1970; Kagermeier, 1981). These species are often added as starters to assure uniform flavour and good conservation (Liepe, 1983). *Lactobacillus sake* and *L. curvatus* dominate at temperatures below 25°C, and above 25°C also *L. plantarum* is found in naturally fermented sausages. *Leuconostoc* spp. and heterofermentative lactobacilli usually represent less than 10% of the LAB flora. The latter are undesirable because of their potential to produce gas, peroxides and slime (Lücke, 1985). Some *L. plantarum* biotypes have been adapted to low temperatures and carry out a specific function in the preparation of sea food. Spoilage is caused by the heterofermentative *L. viridescens*, which turns meat and meat products green (Niven and Evans, 1957). Routine strain identification may be performed by using the differential keys of Schillinger and Lücke (1987).

The pathways of sugar fermentation by these organisms have long been known. An endogenous energy reserve is usually absent. LAB do not possess a respiratory chain and are unable to effect oxidative phosphorylation. The energy-yielding mechanisms are relatively simple, and they rely upon substrate level phosphorylation from sugar fermentations or occasionally on arginine catabolism for the generation of ATP required for biosynthetic and energetic functions of the growing cell (Thompson, 1988). With respect to arginine utilisation, homofermentative LAB are negative, while most heterofermentative LAB and pediococci are positive (Holley et al., 1988). Cysteine can be utilised as an energy source by L. sake L13 in rich medium (Shay et al., 1988). The number of fermentable carbohydrates for lactobacilli is different for various species and decreases in the order L. plantarum > L. sake > L. curvatus. Lactobacillus curvatus is difficult to differentiate from L. sake, since it is not always curved. There is significant DNA homology between L. sake and L. curvatus (Kandler and Weiss, 1986), and they may differ only in the fermentation pattern of a few sugars. Lactobacillus curvatus has a relative low resistance to nitrite and, in contrast to L. sake, hydrogen peroxide (H_2O_2) production is rare (Hastings and Holzapfel, 1987). Nitrate is reduced to nitrite by L. plantarum but not by L. sake and L. curvatus. A typical homolactic fermenter is expected to produce approximately 1.8 moles of lactic acid for each mole of hexose metabolised and about 10% byproducts (Gottschalk, 1979). Deviations from the exclusive production of lactate have been observed during growth of homolactic organisms under specific experimental conditions in continuous cultures at high dilution rates and under nitrogen limitation (Tetlow and Hoover, 1988). Products of heterofermentative metabolism are lactate, acetate, ethanol, diacetyl, acetoin and CO_2. In commercial cheese fermentations with L. plantarum, citrate is often added to increase diacetyl production. Yield coefficients of 73 and 65 μM lactate/mg cell dry weight have been reported for glucose and lactose, respectively. In some strains 20 mM citrate alone may promote diacetyl–acetoin synthesis, and 20 mM pyruvate in even more strains (Montville et al., 1987a,b).

Streptococci (S. diacetylactis, S. lactis) alone or in combination with micrococci or lactobacilli, have been used with success in Russia and former Yugoslavia for the ripening of raw sausages (Schiefer and Schöne, 1978). In eastern German sausages, S. durans has been used as a starter (Scharner, 1990). However, these strains cannot compete at low temperature (Lücke, 1985).

Pediococcus acidilactici (formerly cerevisiae) was the first commercial starter culture for meats, probably because it could easily be lyophilised, reactivated, shipped and stored (Lücke, 1985). Although it is often used for fermentations at elevated temperature (38°C) as in summer sausage, it is a poor competitor at low temperatures (Lücke, 1985). Radiation-killed cells were reported to produce significant amounts of acid at temperatures above 10°C (Rieman *et al.*, 1972). *Pediococcus pentosaceus* is used in the souring of 'Braunschweiger' (Kraft, 1986). There are few data on the basic physiology of pediococci (Tetlow and Hoover, 1988). Generally, they are classified as homofermentative LAB. However, pediococci should not always be considered as producing lactate exclusively under conditions typical for many food fermentations. Amounts of ethanol may be unexpectedly high, as noted in a study on fermentation products from carbohydrate metabolism in *P. pentosaceus* PC39 (Tetlow and Hoover, 1988). *In vitro*, *P. pentosaceus*, *P. acidilactici* and *Lactobacillus plantarum* starter cultures for meat may also produce hydrogen peroxide, with *P. pentosaceus* and *L. plantarum* strains being most active (Juven *et al.*, 1988).

Fermentative activity of *P. pentosaceus* in meat was reported to be affected by penicillin (Raccach *et al.*, 1985). The first report of drug resistance plasmids in *Lactobacillus* was published by Ishiwa and Iwata (1980). In a study by Vidal and Collins-Thompson (1987), 67 strains of LAB isolated from raw ground pork and beef were investigated. The isolates could be separated into four groups, namely atypical streptobacteria, *L. plantarum*, *L. brevis* and *Leuconostoc mesenteroides*. All were sensitive to penicillin, ampicillin, cephaloridin, erythromycin and most strains were sensitive to chloramphenicol and rifampicin. Most strains were resistant against cloxacillin, methicillin, gentamicin, kanamycin, neomycin, streptomycin, tetracycline, polymyxin B, colistin, novobiocin and bacitracin.

Plasmid DNA appears to be widely distributed among the members of the genus *Lactobacillus*. Although most plasmids are cryptic, lactose utilisation, *N*-acetylglucosamine fermentation, antibiotic resistance and bacteriocin production may be plasmid associated (Liu *et al.*, 1988). Maltose plasmids were identified in strains of *L. plantarum* and a *Lactobacillus* spp. isolated from fresh meat. Maltose utilization by *L. plantarum* appeared to be very stable in contrast to the *Lactobacillus* spp. (Liu *et al.*, 1988). Certain sugar degradation pathways are known to be coded on eventually conjugative plasmids, sometimes along with genes for phage resistance or bacteriocin production (Gonzalez and Kunka, 1987; Dunny *et al.*, 1988; Liu *et al.*, 1988). Long-term changes (10 years) in plasmid profile have been observed in *L. plantarum*, however, without loss in phenotype. The starter strain contained seven plasmids from 4 to 19 kbp in size and most of them were quite stable (von Husby and Nes, 1986). Among the LAB, bacteriocins have been described for *Lactobacillus*, *Pediococcus* and *Streptococcus*. Nisin is the most thoroughly characterized bacteriocin among the antimicrobial proteins produced by LAB. Development of bacteriocin overproducers by genetic methods now appears

Table 4.1 Differentiation between *Micrococcus* and *Staphylococcus* (Schleifer, 1986)

Characteristics	*Micrococcus*	*Staphylococcus*
Anaerobic fermentation of glucose	–	+
Resistance to lysostaphin	+	–
Teichoic acid present in cell wall	–	+
MoI% G+C	65–75	30–39

possible and may be an important area of future research in the food industries (Andersson *et al.*, 1988; Klaenhammer, 1988; Mortvedt *et al.*, 1991; Muriana and Klaenhammer, 1991a, b).

4.5.2 Micrococcaceae

In contrast to LAB, members of the family Micrococcaceae usually possess catalase, are acid-sensitive, and more tolerant of low water activities. Strains of the genera *Micrococcus* and *Staphylococcus* are able to multiply and to maintain high cell numbers during ripening of naturally fermented meats. In contrast to micrococci, staphylococci are able to grow anaerobically and to break down glucose fermentatively. Differentiation of micrococci from staphylococci is difficult, since a single enzyme may decide the result of an identification (Weidenfeller and Fegeler, 1990). However, clear differences exist with respect to DNA composition, cell-wall constituents, glucose fermentation and sensitivity towards lysing enzymes (Table 4.1). Both, the non-pathogenic, coagulase-negative staphylococci such as *S. carnosus* and *S. xylosus*, which are often components of starter cultures for dry sausage manufacture, and potentially harmful species like *S. aureus*, may grow on fermenting meats. A strain of *S. xylosus* isolated from salted and dried chub mackerel displayed an a_w limit of 0.91 in glycerol and 0.84 in NaCl–adjusted laboratory media. Generation times at 28°C were 26, 35, 51 and 154 min at a_w values of 0.975, 0.958, 0.943 and 0.919 respectively. Minimum growth temperature was around 3°C. The strain was representative of the dominant microflora in the brine and on the fish (McMeekin *et al.*, 1987; Chandler and McMeekin, 1989).

Micrococci and staphylococci supply two important enzymes in fermented meat production. Catalase may remove hydrogen peroxide from chemical and microbial sources. Nitrate reductase provides nitrite in case curing was with nitrate, and may recycle nitrite that was lost as nitrate in the reaction sequence leading to nitrosomyoglobin. As a result of the chemical degradation of nitrite at pH 5.4–5.5, nitrate and nitric oxide are produced in equimolar amounts. Another source for nitrate is the reaction of oxymyoglobin with nitrite which results in nitrosomyoglobin, nitrate and peroxide. To obtain the full curing colour, about 30–50 ppm nitrite are necessary. Sufficient nitrate reduction occurs in fermented sausage with 10^6 and 10^7 micrococci and staphylococci per gram, respectively. However, when present in sufficient numbers, the bacteria do not have to grow or be viable in sausage mixes to perform their beneficial role in sausage quality (Nurmi, 1966). When applied in the above cell numbers, starter micrococci do not grow during sausage maturation (Holley et al., 1988). Nonetheless, micrococci have been reported to increase initially in numbers during fermentations at temperatures below 24°C reaching maxima of 10^5–10^7/g although weak growth has also been noted. In contrast, micrococci may reach levels of 10^9/g in Italian salami fermented for two days (Holley et al., 1988). This growth behaviour depends largely on the presence or absence of a competitive LAB flora. High initial numbers of LAB may suppress the acid-sensitive micrococcal flora very early in the fermentation process. Since no LAB starters are used in traditional Italian salami, glucose-fermenting staphylococci may easily grow to high levels. Rapid sausage acidulation by rapidly acidifying LAB leads to a rapid decrease in micrococci and may prevent nitrate reduction by the added micrococci (Klettner and Baumgartner, 1980). Acid inhibition of nitrate reduction by naturally occurring micrococci is also observed in treatments with glucono-δ-lactone (GdL). In GdL treatments, micrococci do not grow (Holley et al., 1988). Lactobacilli and micrococci have been reported to reduce nitrite as well, and this may partially explain the low recovery of nitrite from nitrate cured sausages as is typical for this type of product (Lücke, 1985; Holley et al., 1988).

Staphylococcus aureus is still a major cause of food poisoning in fermented meats (Hechelmann et al., 1988). This organism is resistant to salt and nitrite and capable of growth under anaerobic conditions. However, it is a poor competitor at low temperatures, anaerobic conditions and at low pH values. In raw sausages, *Staph. aureus* may grow during the lag phase of lactobacilli development and reach high numbers in the outer layer of the sausages (Lücke, 1985). At least five different enterotoxins may be produced. These are resistant to papain, trypsin, chymotrypsin, are heat stable, but rapidly break down in the presence of reducing agents (ascorbate, nitrite). Effective hygiene measures are necessary to prevent transfer of the organisms from humans to food. Only fresh raw material should be processed, and there should be no interruption of souring with fermented products. An adequate posttreatment (heating/storage) will provide additional safety (Stengel, 1990).

4.6 Occurrence and abundance of meat bacteria

4.6.1 Intestinal habitats

The indigenous mammalian gut microflora is a very complex collection of 10^{14} microorganisms of 400 different types of bacteria (Fuller, 1989). For lactobacilli in pigs it was suggested that 'carbohydrate capsule polymers' are the major determinants of intestinal colonization (Wadström *et al.*, 1987). In addition, lactobacilli and/or streptococci are used as 'probiotics' to supplement animal diets. *Lactobacillus plantarum* is one of many different bacteria claimed to show beneficial effects in animal nutrition. These microorganisms are believed to act as a 'protective flora' by mechanisms of bacterial antagonism and interference towards undesirable microorganisms (e.g. by competing for adhesion sites with pathogens), thus contributing to intestinal balance (Collins and Carter, 1978; Fuller, 1989). The natural killer cell activity is said to be enhanced by *L. plantarum*, and lactobacilli are claimed to prevent the growth of tumours and to be responsible for anticholesterolaemic effects by removing cholesterol from the growth medium.

4.6.2 Raw meat

Microorganisms grow on meat surfaces as discrete colonies (Fuller and Brooker, 1980; Costerton *et al.*, 1985). On prolonged incubation, colonies spread peripherally and tend to coalesce, while still retaining their colony structure. Gram-positive cocci were found to grow in discrete aggregates within a total matrix dominated by Gram-negative bacteria. There appears to be little difference between the pattern of bacterial growth on different kinds of meat (Mattila and Frost, 1988). Lactobacilli frequently found on vacuum-packaged meat include *L. divergens, L. carnis, L. sake* and *L. curvatus* (Schillinger and Lücke, 1987). From irradiated minced beef, all catalase-negative, Gram-positive, rod-shaped bacterial isolates (113) were homofermentative, non-thermophilic, and belonged to the subgenus *Streptobacterium* of *Lactobacillus* (Hastings and Holzapfel, 1987). Nearly all isolates produced DL-lactic acid, were aciduric and relatively insensitive to nitrite and salt. No strain grew at 4°C or was able to reduce nitrate. Pseudocatalase and glycerol fermentation were common. The majority of the strains (88%) were identified as *Lactobacillus sake*. The rest comprised *L. curvatus* (3–8%) and *L. farciminis* (2%). Apparently *L. sake* is less sensitive towards radiation than are the accompanying bacteria, since Niemand and Holzapfel (1984) found only 34% *L. sake* when using half the radiation energy (2.5 kGy). Shaw and Harding (1984) found 57% *L. sake* in non-radurised, vacuum-packaged meats.

4.6.3 Fermented sausage

In general, cell numbers around 10^8–10^9 LAB and 10^5–10^7 micrococci per gram may be found in non-heated fermented sausages. In heated products, the number of LAB varies in sliced dry sausages between 10^2–6.5×10^6g, and lipolytic organisms are present at very different levels of 1 to 2×10^7g

(Giaccone *et al.*, 1988). In fermented sausage, less than 10 coliforms/g meat may be found by day 14 when high-quality meat with a maximum of 2000 coliforms/g is used. This decrease in coliform numbers is typical for most sausage meats, including turkey sausage products (Holley *et al.*, 1988).

4.6.4 Dry-cured ham

Not much is known about the LAB flora of cured ham, possibly because it is normally not very numerous (Hugas and Montfort, 1986). The pH value in these products never falls below 5.7. Consequently it is assumed that the LAB flora cannot have a very significant effect. LAB isolated from ham include *Pediococcus pentosaceus*, homofermentative lactobacilli (*L. alimentarius* (43%), *L. curvatus* (16%) and *L. casei* var. *rhamnosus* (5%) and atypical streptobacteria (26%) and heterofermentative lactobacilli (*L. divergens*). Population levels are not very high ($10–10^4$/g), constituting together with yeasts, the smallest population (Silla *et al.*, 1989). On Spanish dry-cured ham, members of the family Micrococcaceae comprised 90% *Staphylococcus* spp. (56% *xylosus*, 44% *sciuri*) and to 10% *Micrococcus* spp. (Molina *et al.*, 1990). Staphylococci are predominant in this process, probably because of their higher salt tolerance (Seager *et al.*, 1986; Campanini *et al.*, 1987). There are numerous reports on the characterisation of Micrococcaceae from meat products (Comi *et al.*, 1986) which distinguish between two types of staphylococci, the major group being the so-called *Staph. xylosus* biotype II. In German ham, Micrococcaceae consist mostly of coagulase-negative staphylococci (68%) (Rheinbaben and Seipp, 1986), including *Staph. xylosus* and *Staph. sciuri* (Molina *et al.*, 1990). These microorganisms grow in ham during the curing process (15% salt, 3–5°C) but their importance in the biochemistry of ham curing is not known (Molina *et al.*, 1990).

4.7 Meat environment

4.7.1 Meats

Four types of meats are used for raw sausage production, beef and veal, pig meat, sheep meat and poultry meat. Traditionally, most fermented sausages contain pork and/or beef, rarely sheep meat. In addition, fermented sausages may also be prepared from fish (Aryanta *et al.*, 1991). Only beef is used in the typical summer sausage recipe (Burrowes *et al.*, 1986). With respect to poultry, which was used increasingly in the past decade, little published information is available. Raw hams are commonly made out of pork. Normal beef is characterised by a pH of 5.5–5.8 and a glucose content of more than 1.5 mg/g (1.5–3.0 mg/g), while high pH beef has a pH of 6.3–6.8 and a glucose content below 10 μg/g (Dainty *et al.*, 1989b). Dark, firm, dry (DFD) beef may develop putrid odours at a much earlier stage than normal muscle, due to the bacterial attack on amino acids in the glucose-deficient DFD muscles (Newton and Gill, 1978, 1980). Normal pork meat has a pH of 5.6, while soft exudative (PSE) pork has a lower pH (5.4), a reduced water-holding capacity and a paler colour. DFD pork exhibits a higher pH (6.4), greater water-holding capacity, darker colour and a negligible glucose content. The production of pure quality pork has been attributed to genotype and to preslaughter stress. Chronic stress results in the DFD condition of pork muscle while factors such as handling, transit, lairage and mixing may cause acute stress resulting in PSE pork (Greer and Murray, 1988). The inclusion of poultry meats in semi-dry and dry sausages has been the subject of a number of studies. In some of these products, poultry meats were successfully used at very high levels (more than 78%). Two of these studies examined the potential for pathogen growth but sausages were heated at 46 or 60°C, following inoculation and fermentation before drying. In the remainder of the reported work temperatures of 60–74°C were used prior to sausage drying. One of the dangers inherent in the use of poultry meats for sausage manufacture is the threat posed by *Staphylococcus aureus* and *Salmonella* spp. These organisms were found to occur in raw ground turkey meat with varying frequencies, reaching incidences up to 88% for *Salmonella* spp. (Holley *et al.*, 1988). Mechanically recovered meat (MRM) is characterised by a pH of 6.0–7.3 (often 6.7) and a small particle size. Less than 10^6 bacteria per gram are acceptable aerobic plate counts for MRM (Krautil and Tulloch, 1987). Ray *et al.* (1984) have demonstrated that psychro-trophs are able to multiply without any lag phase in MRM at 3°C. *Salmonella* contamination was much higher in MRM than reported in other raw meat and meat products. Spoilage problems may be encountered on thawing pallets of frozen MRM. Therefore, this type of meat should only be used in products that are cooked or canned. It is conceivable that physico-chemical differences may influence the composition of the micro-flora which eventually becomes established. The technological influences of PSE pork on meat products are well known (Wirth, 1986). The use of PSE meat for fermented sausage manufacture was reported to result in slower pH decrease during fermentation, higher weight loss, lower mois-ture, higher protein and fat content, and eventually better texture, flavour and colour. On an industrial scale, PSE pork meat could be mixed with normal meat for the production of fermented dry sausage without quality defects (Honkavaara, 1988). The higher the collagen content of the fat tissue and the lower the amount of unsaturated fatty acids, the better the quality of the end-product. Raw sausages produced with highly unsatur-ated fatty acids may show an unsatisfactory drying behaviour. A notable loss in fat quality has been observed over the last 10 years (Girard *et al.*, 1989). Fermentation is not influenced by the quality of the fat tissue. However, old, rancid fat can cause colour and flavour defects in raw sausages. Apart from mycotoxins, hormones, heavy metals and radionuclides which are undesirable for human consumption, antibiotics directed against bacteria may be present in meats. Generally, at least 80% of all residues in food animals are estimated to originate from the feed (Biehl and Buck, 1987). The polyether antibiotics monensin and lasalocid are routinely fed to feedlot cattle which consume large quantities of cereal grain to prevent ruminal acidosis. These substances bind to the bacterial cell membrane and inhibit the growth of Gram-positive bacteria in a concentration range of 1–20 μM.

4.7.2 Carbohydrates

Glucose can certainly be viewed as the key substrate in the microbiological changes occuring in meat and certain meat products (Nychas *et al.*, 1988). Glucose is also known to be the primary growth substrate for all the common types of spoilage bacteria (Gill and Newton, 1977; Dainty *et al.*, 1989b). Since the glucose concentration in meat varies with meat quality, it is advisable to add 0.1–0.2% of fermentable sugar to the sausage meat. Excess amounts of easily fermentable sugars, however, may result in too sour a product and the development of undesired LAB. The influence of various carbohydrates and bacteria on the fermentation and souring of dry sausages continues to be a field of investigation (Garriga *et al.*, 1988; Liepe *et al.*, 1990). Lactose is cheap and less sweet than glucose or sucrose. It is added in large amounts and often in addition to other sugars for water binding or, in order to obtain economical advantages, as a 'filling' substance in raw sausage technology (Winter, 1988). Lactose is generally not used by LAB in sausage fermentation. Levels in raw ripened dry sausage remained largely unchanged throughout a 28-day ripening period, although the *Lactobacillus* and *Pediococcus* starter cultures applied were able to produce acid from lactose in a laboratory medium. The increase in galactose level (0.04% in 28 days) corresponded to a small decrease in lactose concentration. Again, the above starters were able to ferment galactose *in vitro* (Holley *et al.*, 1988).

4.7.3 Nitrite and nitrate

Current meat-curing practice involves the addition to meat of nitrite, and sometimes nitrate, as well as salt, ascorbate or erythrobate, with or without carbohydrate materials, phosphates and possibly seasonings. Total nitrite and nitrate in fermented raw sausages may vary considerably from about

20 to almost 200 ppm (Arneth and Herold, 1988). In traditional Hungarian salami, a high initial nitrate content (600 ppm) is used (Nagy *et al.*, 1988). However, in recent products, nitrate is omitted and nitrite levels are adapted to the German market. Nitrite is a unique ingredient in meat-curing systems because of its ability to produce the characteristic cured-meat colour and to generate the typical cured-meat flavour. Nitrite possibly acts by stabilising the membrane lipids and in inhibiting pro-oxidants (Ramarathnam *et al.*, 1991). The multiple role of nitrite in the meat-curing process becomes clear from attempts with nitrite-free curing mixtures investigated in cooked meats where nitrite had to be replaced by a complex mixture of substitute chemical compounds (Shahidi *et al.*, 1988). The (putative) reaction product formed between nitrite and ascorbic acid is probably responsible not only for nitrosation reactions, but also for the loss of nitrite observed during curing of meat (Izumi *et al.*, 1989).

4.7.4   Spices

Red pepper has been found to stimulate the rate of lactic acid formation (Vandendriessche *et al.*, 1980), and it was suggested that this may be due to the presence of manganese in the spices (Puglia and Seperich, 1983). Manganese is required by LAB for various enzyme activities including those of the Embden–Meyerhof pathway (Raccach, 1985). The sensitivity of *L. plantarum* strains to nitrite under anaerobic conditions was linked to a dramatic increase in cellular manganese levels. During nitrite inhibition, many homofermentative lactobacilli were found to release CO_2. This release may be the result of manganese increase in the cell and stimulation of the heterofermentative pathway (Collins-Thompson and Thomson, 1986). In bacteria, a manganese-dependent superoxide dismutase, a key enzyme for providing protection against oxidative stress, converts super-oxide O_2^- to hydrogen peroxide (Bowler *et al.*, 1991). The contamination of red pepper used in the making of fermented sausage, with coliform bacteria, including *E. coli*, resulted in spoiled sausages after three days with gas production in the absence of starter cultures. The addition of 0.3% garlic did not show any inhibiting effect (Roca and Kalman, 1989). Ordonez *et al.* (1990) reported a total viable count in pepper of 10^8 cfu/g.

4.7.5   Physical parameters

Several approaches to describe the growth response of microorganisms in foods in relation to temperature, water activity and other rate-limiting factors have been reviewed (Roberts and Jarvis, 1983; Farber, 1986; Baird-Parker and Kilsby, 1987). Modelling microbial growth in response to individual and combined factors is a subject of current research (Chandler and McMeekin, 1989). Good growth of meat bacteria is generally observed at temperatures above 20°C and below 40°C. Sausage fermentations are commonly performed at 22–24°C for 2–4 days, except for summer sausage style-fermentations where higher temperatures are used. Fermentation temperatures as low as 15°C are applied to produce high-quality sausages with long shelf life (Lücke, 1985). For temperatures above 25°C, the risk of faulty fermentations and development of *Staph. aureus* increases in the absence of highly active starter cultures (Raccach, 1981). Ripening (drying) is usually conducted at temperatures below 18°C. Since curing of dry hams is done at 2–6°C, almost no microbial fermentation can be expected (Poma, 1987). Starting water activity for meat batters in fermented sausage production is recommended to be adjusted close to a_w 0.96. Technologically this is basically done by the addition of 2.5–3.0% sodium chloride and about 25–30% fat. With intended reductions in sodium chloride and fat in fermented sausages producers will have to search for suitable a_w-effective substitutes for these components. Gram-negative bacteria may become a problem above a_w 0.97 while *Staph. aureus* may take advantage of the reduced growth of LAB below a_w 0.95. Growth limits for LAB are between a_w 0.91 and 0.94. Therefore, speed of drying is particularly important during the first week of ripening if faults in production are to be avoided (Stiebing and Rödel, 1988).

Oxygen has a pronounced effect on the growth and metabolism of LAB. Many LAB grow well aerobically and the aerobic growth rate may even be higher than the anaerobic rate. The presence of oxygen may also increase the yield of biomass. However, for some LAB, air is toxic, due to the accumulation of toxic oxygen compounds that affect the ability to grow (Condon, 1987). In the presence of oxygen, metabolites other than those found in anaerobic cultures may be observed. Hydrogen peroxide may be synthesised during the aerobic metabolism of glucose and the product yields of lactic acid, acetic acid, acetoin and ethanol are affected. *L. plantarum*, which under anaerobic conditions mainly forms lactic acid from glucose, shows a dramatic increase in the production of acetic acid under aerobic conditions, together with a minor production of acetoin (Murphy and Condon, 1984; Sedewitz *et al.*, 1984). For heterofermentative *Lactobacillus* spp. in the presence of oxygen, the yield of acetic acid may increase at the expense of ethanol and lactic acid (Blickstad and Molin, 1984). In batch cultures at pH 6.0, 25°C and 1% glucose, homofermentative lactobacilli show a maximal oxygen consumption during the stationary growth phase (Borch and Molin, 1989).

Leuconostoc strains, heterofermentative lactobacilli, *Brochothrix thermosphacta* and *Carnobacterium* strains showed a maximal oxygen consumption during the exponential growth phase. The maximum specific growth rate varied from 0.19 to 0.54 per hour while the growth yield varied from

19 to 86 g dry weight per mole glucose consumed. In general, homofermentative lactobacilli produce DL-lactic acid, acetic acid and acetoin. Heterofermentative lactobacilli produce DL-lactic acid, acetic acid, and some strains also produce ethanol. All three compounds are also formed by *Leuconostoc* species. *Brochothrix thermosphacta* produces acetoin, acetic acid, formic acid, isobutyric acid and isovaleric acid but no lactic acid. Apparently there is no measurable oxygen present in raw meats and finished meat products, including long ripened raw sausages. The partial pressure of oxygen (pO_2) in slaughter-fresh pork was reported to be basically zero (Rödel *et al.*, 1990). In raw-sausage batter prepared with nitrite curing salt, where oxygen is introduced during comminution, a pO_2 (at 2°C) of 14 and 7 mmHg was measured 2 and 6 hours after production, respectively, Nitrate-cured meat reached zero-oxygen levels much faster compared to nitrite-cured meat. This might be due to inhibition of oxygen-consuming microorganisms by nitrite, but probably even more to general interference with possible reactions of molecular oxygen within the batter. No measurable oxygen was present in the batter after 24 hours.

4.8 Sausage ripening

The essential role of lactic acid in food technology is the inhibition of growth of food spoilage bacteria. In addition, both stereoisomers of lactic acid have an acceptable taste and are easily metabolised by the human body (Stackebrandt and Teuber, 1988). A general guide to the manufacture of fermented sausages is given by Lücke (1985). The ripening process of fermented sausage depends considerably on the changes brought about by bacteria. In 'aseptically' manufactured dry fermented sausages, containing less than 100 microorganisms/g initially, less than 10^5 bacteria/g were present after 20 days of ripening. In the absence of micrococci no reduction of nitrate to nitrite occurred and the sausages had the typical brown colour of metmyoglobin (Ordonez *et al.*, 1990). Batches inoculated only with micrococci developed the usual numbers (about 10^6/g) of these organisms, whilst numbers of lactobacilli were about 10^5/g at the end of the ripening.

A uniform distribution of microorganisms throughout the sausage was reported by Liepe (1987). However, by using electron microscopy, Katsaras and Leistner (1988) found a clustered distribution of microorganisms in salami, the nests being 0.1–5 mm apart from each other. *L. plantarum* and *Staph. aureus* were reported to occur in higher numbers in the outer layer of the sausage as compared to the sausage core. It was reasoned that oxygen availability might be responsible for this behaviour (Lücke, 1985).

In North America, lactobacilli or pediococci are used as starter cultures either separately, together or combined with micrococci or coagulase-negative staphylococci (Bacus, 1984). In Europe, micrococci/staphylococci have been used alone or combined with lactobacilli and less use has been made of pediococci. However, pediococci are becoming more frequently observed in starter culture materials since the description of low temperature strains of *P. pentosaceus* (Holley *et al.*, 1988). At present, no satisfactory LAB-starter cultures for maturation of the delicately flavoured, low-acid, Italian raw-ripened sausages are available on the market (Cantoni and Bersani, 1985). Yeasts (*Debaryomyces* spp.) have also been used with LAB for dry-sausage ripening and a mould starter culture for surface application has been available for some time. *Streptomyces griseus* is used as a starter culture in combination with *Staph. carnosus* in Finland (Honkavaara, 1988). Additionally, Gram-negative bacteria have been used with limited success in these processes.

The amount of sodium chloride used in most sausage products inhibits the majority of fresh-meat spoilage bacteria, i.e. *Pseudomonas* spp. and Enterobacteriaceae, and favours the growth of micrococci, streptococci and lactobacilli (Kraft, 1986). The number of lipolytic microorganisms increases during ripening to maximum levels of 10^5 organisms/g and is thought to be responsible for the increase in free lipid-acid content (Nagy *et al.*, 1988). Factors which favour the growth of Enterobacteriaceae during sausage fermentation, and thereby increase the risk of *Escherichia coli* and *Salmonella* spp., include a high initial water activity, a high initial pH, a low concentration of fermentable carbohydrates, a low number of lactobacilli in the fresh sausage mixture, the use of nitrate or very low levels of nitrite as the curing agent and high ripening temperatures (Leistner *et al.*, 1973; Lücke, 1985).

The behaviour of LAB during sausage fermentation may be outlined for a strain of *Lactobacillus curvatus* (Bantleon, 1987; Gehlen, 1989). The isolate was found to be very competitive against the spontaneous microbial flora, showed a fast and sufficient fermentation characteristic and had a favourable influence on sensoric properties in raw fermented sausage cured with nitrite at 20–24°C. A less homofermentative fermentation was observed when higher initial ripening temperatures were applied. In the presence of 0.3% glucose the pH dropped to 5.2–5.4 within 48 hours and reached a constant value of 4.9–5.0 after another 3–5 days. At this time, the sausage was firm enough for slicing and showed an a_w of 0.93–0.94. The lag phase may be up to 24 hours, after 2–5 days 10^9 organisms/g are reached, and the spontaneous flora is mostly suppressed. A minimum inoculation density of 10^6 cells/g is necessary to avoid an increase in pH. The strain will readily grow even with an inoculum of less than 10^5 cells/g, but poor sensory properties were observed in this case. With 2×10^6 organisms/g, the pH increased until the second day, due to a decrease in lactic acid content. The outside colour of the sausages was improved. However, with large-diameter sausages (80 mm), the decrease in pH may occur too slowly. In this context, *L. plantarum* was discussed as being unreliable, since, in contrast to the *L. curvatus* strain, it was not the dominant LAB during the whole fermentation and ripening process. Since it has a lactose-negative phenotype, the *L. curvatus* strain has some potential for use in sausages made with high amounts of added lactose. However, with lactose as the only added sugar acidification and nitrate reduction were reduced. The lactic acid was degraded in the presence of oxygen to acetic acid and carbon dioxide, resulting in a porous product with a biting taste. Only final pH but not microbial growth was influenced by the amounts of sugar added. The addition of more glucose was suggested to result in better structural properties and better aroma of the sausages. Sausages which were too soft were obtained when only 2% NaCl was used. In the absence of starter micrococci, nitrate was reduced by spontaneously growing staphylococci. Curing with both nitrate and nitrite, resulted in rancid products with no development of the curing colour in the core of the sausages, the reason being a reduced nitrate reduction due to inhibition of the spontaneous staphylococci.

In 'summer sausage' with a 1% inoculum of *L. plantarum* (2.5×10^6 cells/g sausage mixture) the pH of the sausages drops from 5.8 to 4.9 within 16 hours (Burrowes *et al.*, 1986). The increase in cured meat pigment content during fermentation is the result of two factors: (i) the cumulative heat energy input and (ii) the decrease in sausage pH (Demasi *et al.*, 1989). The nitrosohaeme content increased during processing of summer sausage from 12% of the total haeme pigment, immediately after mixing at 8°C and pH 6.0, to 71% after 9.7 hours of fermentation at 38°C (pH 4.65). Light exposure, as compared to dark storage, can affect all colour properties and reduces the nitrosohaeme pigment content of fermented sausages (Demasi *et al.*, 1989).

Rancid products result from lipid oxidation in the sausages. The relative roles of inorganic iron and haemoproteins as catalysts of lipid oxidation in meat are still a subject of much debate (Asghar *et al.*, 1988; Rhee, 1988; Johns *et al.*, 1989). Haemoglobin (10–100 μM) is a very powerful catalyst of lipid oxidation in contrast to inorganic iron (Johns *et al.*, 1989). Recent work, primarily by Harel and Kanner (1985a,b) and summarised by Rhee (1988), suggests that ferric haeme pigments may only be effective catalysts in the presence of hydrogen peroxide (Johns *et al.*, 1989).

Slow ripened dried sausages are stable at room temperature for many months. Under optimum conditions it is possible to store sliced, vacuum-packaged salami for up to 41 days at 6°C (Wiegner and Hildebrandt, 1986). The minimum shelf life of sliced, vacuum-packaged Italian dry sausage is 28 days (Giaccone *et al.*, 1988). Fermented meats that are not subjected to a drying process after fermentation, like *nham* or German spreadable raw sausages, are usually not shelf stable and are best consumed one or two days after production.

4.9 Control of fermentation

Low-temperature holding of meat mixes in bins or trays with or without added starter and curing ingredients prior to stuffing was at one time, common practice, especially prior to the adoption of bacterial starter cultures (Holley *et al.*, 1988). The usefulness of this practice has been disputed. Longer holding (12 days at 5°C) did facilitate the growth of naturally occurring lactobacilli in salted, mechanically separated chicken meat (MSCM). A one-day pre-incubation of inoculated MSCM, containing 2.7% NaCl and 1% glucose, at 10°C was not useful according to Holley *et al.* (1988). Similarly, no advantages of pre-incubation were found by Deibel *et al.* (1961) and Joseph *et al.* (1978), who also pre-incubated starter-inoculated meat. With Italian-type salami, Rödel *et al.* (1991) observed a reduction in the overall bacterial flora.

In order to achieve a better control of the fermentation process in the presence of starter cultures, Landvogt and Fischer (1990) calculated the acidification rates (pH/hour) for two strains of *L. curvatus*, for *P. pentosaceus* and for *P. acidilactici*. The calculations were done from acidification curves obtained in a model sausage meat for different water activities and temperatures. Under normal conditions, the time span for *L. curvatus* to reach pH 5.2 increased from 17 to 210 hours when the temperature was decreased from 30° to 10°C. With decreasing water activity, the lag phase increased and a higher final pH resulted. *P. pentosaceus* and *P. acidilactici* gave comparable results. *L. curvatus* was very similar but one strain was much more active at lower a_w values than the pediococci. At higher temperatures, the rates were more affected by decreasing a_w and the type of strain used. The a_w minimum for acidification was around 0.93 for the spontaneous flora, 0.92 for the pure cultures and 0.91 for one strain of *L. curvatus*. The acidification rates at these a_w values were found to be independent of temperature. Higher temperatures increased the acidification rates of the pure cultures more than those of the spontaneous flora. The degree of comminution and glucose concentrations in excess of 0.15 %, did not influence the acidification rates, but higher bacterial inocula did. The acidification rates of the pediococci were comparable to those of the spontaneous flora and were half those of the lactobacilli. Without glucose, only a 0.1–0.3 pH reduction was obtained (Landvogt and Fischer, 1991).

4.10 Fermentation products

Besides lactic acid, the main fermentation product in sausage manufacture a variety of other chemical compounds may be found in the end-product in varying, but usually small concentrations. These compounds may influence the final quality of fermented meats either negatively or positively. Although many of these compounds may be formed by microorganisms *in vitro*, their microbial origin and significance in fermented meats is not always clear. Similarly, the qualitative and quantitative contribution of individual microorganisms, indigenous meat enzymes and non-enzymatic reactions to the final concentration of certain compounds is basically unknown. Gas production (CO_2, H_2S) by microorganisms is generally undesirable in raw-sausage production, as are excess amounts of ethanol, acetic acid and biogenic amines. The production of hydrogen peroxide and hydrogen sulphide may lead to significant colour and flavour defects. On the other hand, a certain degree of proteolysis and lipolysis may improve overall flavour of fermented meats (Lücke, 1985).

4.10.1   Acids

Lactic and acetic acids are the major fermentation products in raw sausages. In the absence of added starter cultures, lactic acid formation is due to *Lactobacillus* spp. (*L. sake, L. curvatus, L. plantarum*). Most fermented sausages in Belgium and Germany have pH values of 4.8–5.0 after fermentation, corresponding to a concentration of 25 g of lactic acid/ kg dry weight (Lücke, 1985). The molar ratio of lactate to acetate is between 7 and 12, and an initial glucose concentration of 1.5% may be estimated from these data. Butyric and propionic acids are about one thousand times less in concentration (50–200 μmol/kg dry weight). For long ripened products with a final pH of 5.1 after 42 days, Girard *et al.* (1989) reported 70–110 μmol lactate and 25–50 μmol acetate per gram dry weight. This would correspond to 8.6–14.4 mg/g dry weight of glucose metabolised or at least 5.7–9.6 mg/g (0.6–1.0%) present initially. Both acids contribute, in their undissociated form, to inhibition of spoilage bacteria. Acetic acid exerts its greatest effect on total aerobic counts, followed by Enterobacteriaceae, with *E. coli* being least affected (Anderson and Marshall, 1989).

4.10.2   Hydrogen peroxide

Oxy-radicals are byproducts of many biological oxidations. The toxicity of O_2^- radicals and H_2O_2 *in vivo* arises by a metal ion-dependent conversion into hydroxyl radicals. Of particular importance is its ability to mutate DNA and to initiate chain reactions of lipid peroxidation. Sources are the electron-transport chains of mitochondria and microorganisms, but also lactate catabolism by LAB in the presence of oxygen in addition to their generation from these indigenous reactions, oxy-radical formation appears to be greatly increased during stress conditions (Juven *et al.*, 1988; Bowler *et al.*, 1991). Hydrogen peroxide is one of the primary metabolites that may be produced by LAB and which may contribute to their antagonistic action. The production of hydrogen peroxide by LAB varies between strains and is dependent on the availability of oxygen (Collins and Aramaki, 1980). Maximum levels of H_2O_2 production *in vitro* were reported to be 0.85 μg $H_2O_2/10^9$ cells (22 nmol) for the investigated meat starter cultures *P. cerevisiae* and *L. plantarum* (Raccach and Baker, 1978). Lücke *et al.* (1986) reported 60 nmol H_2O_2/min/mg protein with DL-lactic acid as an electron donor for washed cells of aerobically grown *Lactobacillus*. The peroxide-forming enzymes are fully induced only in the presence of oxygen. Many strains of the *L. sake/curvatus* group produce significant amounts of H_2O_2 (Lücke *et al.*, 1986; Hastings and Holzapfel, 1987), and production rates may even be higher than with *L. viridescens* (Lücke *et al.*, 1986). Knowledge of H_2O_2 formation in fermented meat products is of importance in relation to explaining antimicrobial effects, colour and flavour changes. However, once formed, H_2O_2 may react with chemical constituents of the growth environment and therefore be undetected by analytical methods (Juven *et al.*, 1988). During lipid oxidation in meats, one of the major causes of quality deterioration, H_2O_2 activates metmyoglobin, probably to a porphyrin cation radical which initiates lipid peroxidation (Harel and Kanner, 1985b). Membrane lipid peroxidation may be initiated in the presence of 30 nmol metmyoglobin preactivated with 30 nmol of H_2O_2 (Juven *et al.*, 1988).

4.11 Biogenic amines

Biogenic amines result from amino acid decarboxylation by microorganisms. Their presence in foods at higher levels may pose a health risk to the consumer (Table 4.2). Maximum levels of about 30 and 60 μg/g surface tissue of putrescine and cadaverine, respectively, were reported for vacuum-packaged beef inoculated with $10^3–10^4$ organisms/g and stored for 8–9 weeks at 1° C (Dainty *et al.*, 1986). Of a variety of meat strains of streptobacteria, leuconostocs, Enterobacteriaceae, pseudomonads and

Table 4.2 Negative aspects of biogenic amines (Edwards *et al.*, 1987b)

Histamine	Vasoactive properties, scombrotoxin poisoning
Tyramine	Vasoactive properties, migraine attacks, hypertensive crisis in patients being treated with monoamine oxidase inhibitors as antidepressants
Spermine spermidine	Contain a secondary amine group, therefore may form *N*-nitrosamines with nitrite
Putrescine cadaverine	May be converted by heating into compounds that may form *N*-nitrosamines with nitrite
Trimethylamine	Highly undesirable sensory properties

Brochothrix thermosphacta tested, only *Hafnia alvei* and *Serratia lique-faciens* showed diamine-producing potential during growth in pure culture on beef stored in vacuum packs at 1° C (Dainty *et al.*, 1986). Two groups of atypical streptobacteria produce ammonia from arginine, suggesting the presence of arginine dihydrolase, an enzyme found in many other LAB. Because this enzyme produces ornithine as an end-product (Abdelal, 1979) and several strains of Enterobacteriaceae produce ornithine decarboxy-lase, putrescine could be formed by combined activities of the two bacterial types. In Enterobacteriaceae, putrescine formation from arginine, through the combined activities of arginine decarboxylase and agmatine ureohydro-lase, is a well-established pathway (Abdelal, 1979) but activity of the latter is weak at pH 5.5 (Goldschmidt and Lockhart, 1971). Even if the Enterobacteriaceae are eliminated during extended storage, the diamines will be indicative of their former presence and potential growth (Dainty *et al.*, 1986). Many *Pseudomonas* spp. are able to produce ornithine from arginine via arginine dihydrolase. Some strains can then produce putres-cine from ornithine through decarboxylation. *P. fluorescens*, *P. fragi* and *Alteromonas putrefaciens* form putrescine in pork stored aerobically at 5–8° C (Slemr, 1981), possibly by using the same pathway as indicated for Enterobacteriaceae. In microbes, the usual precursor for putrescine is arginine (Abdelal, 1979). *Lactobacillus brevis*, a beta-bacterium growing in beer wort (Zee *et al.*, 1981) and an oral *Lactobacillus* (Lagerborg and Clapper, 1952), formed putrescine when incubated with ornithine.

Fresh beef of normal pH contains putrescine, cadaverine, histamine, spermine and spermidine. Development of a natural spoilage flora during vacuum-packaged storage at 1° C led to increases in concentrations of putrescine and cadaverine, and to the production of tyramine. Tyramine formation was restricted to lactobacilli (*Lactobacillus divergens* and *L. carnis*). Strains of *Leuconostoc*, Enterobacteriaceae, *Pseudomonas* spp. and *Brochothrix thermosphacta* were negative. Production of tyramine occurred at cell densities of $<10^6/cm^2$ (Edwards *et al.*, 1987b). Measurable increases in cadaverine concentration were found in meat stored vacuum-packaged at 1° C for 2–3 weeks when still low cell numbers of the responsible organisms, *Hafnia alvei* (27 µg/g surface tissue, 3.2×10^6 organisms/cm²) and *Serratia liquefaciens* (8 µg/g surface tissue, 2.0×10^4 organisms/cm²) were present. But at least 10^6 organisms/g were necessary in the case of putrescine (Dainty *et al.*, 1986). There is evidence that production of tyramine by lactobacilli occurred at cell densities of $<10^6/cm^2$. At $10^6/cm^2$, there were no overt sensory signs of microbial growth. Increases of putrescine and cadaverine were evident after five weeks of storage when bacterial numbers were approaching, or in excess of, $10^7/cm^2$. However, the biogenic amine levels observed in this study were no cause for concern. Putrescine and cadaverine formation is dependent on the growth of Enterobacteriaceae to substantial numbers (Edwards *et al.*,

1987b). Enterobacteriaceae strains were shown to be a major source of cadaverine during vacuum-packaged storage, while the combined activities of LAB and Enterobacteriaceae strains were required for putrescine formation at the levels detected in naturally contaminated meat (Dainty *et al.*, 1986). If Enterobacteriaceae are omitted, putrescine will not be formed in vacuum-packaged meats. LAB are also assumed to be the source of the tyramine frequently detected in a range of dry and semi-dry fermented sausages (Santos-Buelga *et al.*, 1986), its formation being primarily associated with the short, initial fermentation stage rather than the subsequent drying and maturation phase. The amounts recorded in these products can exceed 100 μg/g and are therefore approaching levels considered to be of some concern to individuals susceptible to tyramine (Rice *et al.*, 1975).

4.12 Aroma compounds

The literature available on cured meat flavour is very limited. Meat-flavour volatiles have been reviewed and discussed by Ramarathnam *et al.* (1991) with respect to lipid oxidation of the original meat (pork, beef, lamb, poultry). Raw meat itself has little odour and only a blood-like taste, whereas cooking and fermentation develop the flavour. Nitrite curing remarkably simplifies the flavour spectrum of meat (Ramarathnam *et al.*, 1991). Cured-meat flavour comprises the basic meat-flavour compounds derived from non-triglyceride precursors. Species differences are due to differences in the spectrum of carbonyl compounds. More than 50 hydrocarbons, 37 carbonyls, six acids, two alcohols, but no nitrogenous or sulphur compounds were reported in a recent study where the volatile carbonyls and hydrocarbons of uncured and cured cooked pork were compared (Ramarathnam *et al.*, 1991). In both cases, carbonyl compounds were found in large amounts. Hexanal, a major lipid oxidation product, was found to be present in uncured meat at a concentration of about 13 ppm, while only 0.03 ppm was present in the cured product. Also, the concentration of other carbonyl compounds was higher in uncured pork, while they were either present in reduced amounts or not detectable in the cured meat. Although a variety of factors are known to influence the flavour of meats, no single group of factors can be assigned the principal role. The volatiles of cooked cured and uncured ham are qualitatively similar but quantitatively different. Striking differences exist in *n*-pentanal and *n*-hexanal concentrations.

Although the nature of cured-meat flavour seems to be much simpler than that of uncured meat, and is postulated to be the basic flavour of meat regardless of species, identification of the compounds responsible is difficult. In addition, minute traces of compounds can be aroma effective

Table 4.3 Compounds related to meat spoiling pseudomonads (Edwards *et al.*, 1987a)

Odours	Compounds	*Pseudomonas*
Cabbagey, sulphury, putrid	Methanediol and its derivatives	spp.
Fruity	Ethyl esters of short-chain fatty acids	*fragi*
Strong garlic	Isopropanethiol and its derivatives	spp.
Cheesey, dairy	Methyl ketones with an odd number of carbon atoms	*fragi, putida, fluorescens*
Floral, alcoholic	2-Heptyl acetate, 2-heptanol, 2-heptanone	spp.

(Ramarathnam *et al.*, 1991). Variations in meat composition (substrate availability), non-enzymatic processes such as lipid oxidation, reactions catalysed by meat and/or bacterial enzymes and analytical methodology may all contribute to the number of aroma compounds detected (Edwards *et al.*, 1987a). To date, no isopropyl compounds have been reported in naturally contaminated meat samples, but methanethiol and derived sulphides have, and they are believed to be important components of the sulphury, putrid, cabbagey odours which develop (Table 4.3). Microbial metabolism of amino acids and/or carbohydrates may result in the release of thiol and ester compounds. Fatty acids probably derive from lipids. However, their formation via non-microbial activities cannot be ruled out. Autolytic and/or oxidative breakdown of lipid may result in alcohols, hydrocarbons, alkanals and methyl ketones with six or more carbon atoms. (Edwards *et al.*, 1987a). Volatile compounds of major sensory significance, produced by meat isolates of *P. fragi* and *P. fluorescens* biotype 1 during growth on beef at 6°C in air, were ethyl and methyl esters of C_2–C_8 fatty acids and sulphur-containing compounds. These included methane and isopropane thiols and their related sulphides and thioesters but not hydrogen sulphide. Ethyl esters were invariably present in greater numbers and total concentration than methyl esters. *Pseudomonas fragi* may produce one or more of aroma-active methyl ketones with an odd number of carbon atoms (Edwards *et al.*, 1987a). The absence of acetoin can be attributed to the known ability of pseudomonads to metabolise the compound (Molin and Ternström, 1986; Edwards *et al.*, 1987a).

Acetaldehyde and ethanol, are typical flavour compounds that may be found in sausage fermentation. The primary volatile flavour metabolite produced by *L. plantarum* in summer sausage was acetaldehyde in amounts of 15 and 19 nmoles/g (0.7–0.8 ppm) after 16 and 72 hours, respectively (Burrowes *et al.*, 1986). When *Leuconostoc* and *L. plantarum* were inoculated at 10^6 cells/g each, ethanol was identified as the major flavour compound (1.2 μmoles/g or 55 ppm). Diacetyl was produced in small amounts (less than 0.1 ppm). The more tangy flavour of acetalde-hyde was preferred over the more mellow spicy flavour, in agreement with traditional expectations of consumers towards more tangy sausage. At the time of sensory evaluation, 33 and 18 nmoles of acetaldehyde were detected in each sausage type, respectively. In comparison with the acetaldehyde levels, 40–60 times higher concentrations of ethanol were found in the presence of *Leuconostoc*. On the basis of 11 μmoles of lactic acid formed from 0.1% glucose, ethanol production would correspond at maximum to 10% of the lactic acid, and maximum acetaldehyde production to 0.3%.

4.13 Important enzymes

4.13.1 Nitrate reductase

Reduction of nitrate to nitrite is found with *Staph. xylosus, Staph. carnosus, Staph. simulans, Staph. sciuri* and *Micrococcus varians. Staphylococcus carnosus* may further reduce nitrite to ammonia (Schleifer, 1986). *Lactobacillus plantarum* can reduce nitrate *in vitro* but not under the conditions of raw sausage fermentation (Lücke, 1985). As mentioned already, less than 50 ppm of nitrite is sufficient to obtain the full curing colour in fermented sausages. This amount has to be reached before significant pH reduction as a result of LAB activity since bacterial nitrate reduction becomes insignificant at pH values below 5.4. Because of the wide variability of fermentation conditions (temperature, pH, presence/absence of lactic and/or micrococcal starter cultures) it is difficult to postulate a specific enzyme activity (ppm nitrite/hour/number of organisms). According to Gehlen (1989), 10^7 cells of *S. carnosus* (Sc76) per gram sausage reduce about 125 and 200 ppm KNO_3/day (equivalent to 84 and 135 ppm $NaNO_2$ respectively) at 18°C and 24°C, respectively. *Micrococcus varians* M28 showed about 50% less activity, Nitrate reduction rates were almost constant in the first 5–6 days of fermentation and then rapidly declined. Therefore, at least 10^6 micrococci per gram of sausage are necessary when nitrate is used as a curing agent. No attempts have been made to relate *in vitro* activities of microcci to activities found in fermenting meats. In the curing process of dry hams, low temperature in concert with high salt concentration reduces nitrate reduction by a factor of 10–100 compared to raw sausage ripening.

4.13.2 Catalase

Since hydrogen peroxide may be released by LAB, catalase provided by micrococci could prevent colour and flavour defects in fermented sausages (Lücke, 1985). Catalase is present in bacteria that employ a respiratory cytochrome electron transport system for energy generation, i.e, in all aerobes and most facultative anaerobes. Catalase is not generally present in strict anaerobes or the LAB, although some strains can contain true or pseudocatalase activity (Kroll *et al.*, 1989). Micrococcal starter cultures, added in the usual cell numbers to raw sausage batter (10^6–10^7 cells/g) could remove around 15 nmoles H_2O_2/min, according to data from Kroll *et al.*, (1989). As low as 10^3 bacteria/ml can be detected *in vitro* when assessing their catalase activity. In foods, the sensitivity is reduced to about 10^5 bacteria/ml due to non-microbial catalase (Kroll *et al.*, 1989). With minced beef and braising steak, significant amounts of indigenous catalase may be present, masking in most cases any bacterial catalase. There are large differences between the catalase activity of different organisms, with the Gram-positive bacteria having significantly more catalase activity per cell. In pure cultures, the specific activity of *Micrococcus* and *Bacillus* was about four times higher than that of *Pseudomonas* and *E. coli*. The catalase content of bacteria is dependent upon growth phase and growth conditions (Hassan and Fridovich, 1978; Loewen and Switala, 1987; Martin and Chaven, 1987). In *Staphylococcus aureus*, catalase activity increases at the onset of the stationary phase (Martin and Chaven, 1987).

4.13.3 Lipase

Lactobacilli and *Brochothrix thermosphacta* have frequently been reported as the dominant microflora on meat and meat products and may spoil them by producing souring (Egan, 1983) or a pungent cheesy flavour (Gardner, 1981). Off-flavour development and spoilage in meats, also subtle flavour changes characteristic of certain aged meats can all be attributed to microbial lipases (Alford *et al.*, 1964). Despite the possible participation of LAB lipases in the flavour changes in meats, little information is available on their lipolytic activity (Papon and Talon, 1988). Lipase production by microorganisms depends greatly upon the growth phase, the incubation conditions and the composition of the medium. Endogenous lipases may be released in small amounts upon cell lysis. Most bacterial lipases have their activity at a neutral pH (Papon and Talon, 1989). Among lipolytic bacteria found in meat only certain Micrococcaceae and *P. fragi* have been shown to be lipolytic against natural pork fat (Cantoni *et al.*, 1967; Debevere *et al.*, 1976). Lipolytic activity of lactobacilli strains and *Brochothrix thermosphacta* is cell-related. No significant activity was found in the supernatant fluids. The lipases of *Brochothrix thermosphacta* and *Lactobacillus curvatus* were found in the soluble fraction of the cell and were not membrane- or wall-bound. Their optimal activities (20 µEq fatty acids/hour/mg protein) were around pH 6.5 and at 37°C. Maximal activity was found with tributyrin and hydrolysis of other triglycerides decreased as the length of fatty acid increased. At a chain length of 10 carbon atoms, only 11% of the C$_4$-activity was found. The lipases of *Brochothrix thermosphacta* and *Lactobacillus curvatus* do not belong to the group of markedly thermostable lipases found in *P. fragi*. They compare best with *Propionibacterium shermanii* and *Streptococcus faecalis* whose lipases are also inactivated by heating.

Brochothrix thermosphacta and *Lactobacillus curvatus* synthesise lipases at low temperature. These lipases might be active under conditions used for cold storage of food, although the role of these enzymes in the spoilage of fatty foods would be limited (Papon and Talon, 1989). Most lipase was produced during the logarithmic phase of growth and was greatly affected by growth conditions. For all strains, an initial pH around 7.0 and low glucose concentrations stimulated lipase production. Butyric acid and anaerobic conditions inhibited lipase production by *Brochothrix thermosphacta* but not for lactobacilli (Papon and Talon, 1988). Lipase production by *Acinetobacter* spp., *P. fragi* and *P. aeruginosa* is maximal during the logarithmic phase of growth and then decreases. With LAB, activity decreases with decreasing pH of the medium. Lipase activities of lactobacilli and *Brochothrix thermosphacta* are still important at pH 5.6, the normal pH of meat. At 3–7.5°C, lipase production remains significant (Papon and Talon, 1988). Lactobacilli (Sanz *et al.*, 1988) and lipolytic micrococci (Selgas *et al.*, 1988) are able to hydrolyse short-chain fatty acid glycerides. A lipolytic *Micrococcus* was shown to produce higher amounts of short-chain fatty acids than a non-lipolytic species in an aseptically reinoculated sausage meat (Ordonez *et al.*, 1990). Higher levels of acetic acid may also result from sugar or alanine catabolism by lactobacilli (Dainty *et al.*, 1986). Lipolytic staphylococci include strains of *Staphylococcus simulans, Staph. xylosus* and *Staph. sciuri*, the last showing only a weak lipase activity (Schleifer, 1986). The addition of lipase, isolated from LAB, during salami fermentation produced a bitter taste and increased stickiness and meat whiteness in the final product (Naes *et al.*, 1991).

4.13.4 Protease

Proteases have been used widely in the food industry. Lactococci growing in milk are dependent on their proteinase enzyme systems and efforts are being made to genetically improve these systems in starter cultures used for cheese ripening (McKay, 1986). Proteolytic processes also contribute significantly to both an improved final meat aroma and spoilage of meats. Amino acids and their degradation products may contribute to the taste of fermented meats. From the older literature, it appears that during the ripening of fermented sausages a significant portion of the protein is converted into non-protein nitrogen, mainly by microbial proteases (Lücke, 1985). At present, however, it seems unclear to what extent LAB and/or micrococci participate in proteolytic changes during the ripening of raw sausages and raw hams. Data on the presence of proteinases in lactobacilli, pediococci and leuconostocs are very scarce (Kok and Venema, 1988). LAB are weakly proteolytic compared with many other groups of bacteria (*Bacillus, Proteus, Pseudomonas*, coliforms). Lactobacilli probably have cell wall-bound proteinases. However, there is no direct evidence that these organisms contribute significantly to fermented meat flavour and it is doubtful whether these enzymes play any role at all in meat proteolysis (Law and Kolstad, 1983). Proteinase extracted from *L. casei* had some effect on hardness of salami during sausage fermentation (Naes *et al.*, 1991). Of the micrococci, *Micrococcus varians* and *Staphylococcus sciuri* are able to hydrolyse gelatine, while *Staph. xylosus* and *Staph. carnosus* are not (Schleifer, 1986). Significant proteolytic activities are mostly due to the Gram-negative spoilage flora, although moulds used in ripening processes of southern European fermented meats are also able to secrete proteinases. Moulds are the most likely candidates for flavour improvements of raw sausages and hams due to hydrolysis of meat proteins. Besides, meat proteolytic enzymes themselves may be active not only during ripening of raw ham but also during sausage fermentation.

4.14 Spoilage

Changes in traditional recipies, especially reductions in sodium chloride, nitrite and fat may result in unstable products, ultimately leading to food spoilage and/or food poisoning. The first hours and days (high a_w and pH) are critical with respect to *Salmonella, Listeria* and *Staph. aureus*. Stable products should soon reach a pH below 5.4 (Hechelmann and Kasprowiak, 1991). *Pseudomonas* spp. and *Brochothrix thermosphacta* are common spoilage organisms of fresh meat (Gardner, 1981; Gill, 1983). Pseudomonads typically dominate the flora of meat stored in air at chill temperatures (Molin and Ternström, 1986; Edwards *et al.*, 1987a,b; Shaw and Latty, 1988;). Once they have exhausted the glucose available in meat pseudomonads will metabolise proteins. Deamination of proteins by the bacteria produces ammonia, presumably the cause of the observed pH increase (Gill, 1983; Dainty *et al.*, 1986). Marriott *et al.* (1967) identified *P. fragi* and *P. geniculata* as the predominant bacteria on prepackaged fresh beef steaks. A considerable amount of research has been conducted on the spoilage effects of *P. fragi* on meat products (Molin, 1985). *Brochothrix thermosphacta* produces mixtures of acetoin, diacetyl, and four- and five-carbon branched-chain acids, aldehydes and alcohols, resulting in creamy, dairy, and cheesy odours (Gardner, 1981; Dainty *et al.*, 1989b). Growth was observed on glutamate (Gill and Newton, 1977) although earlier reports found *Brochothrix thermosphacta* being exclusively dependent on saccharolytic processes (Sneath and Jones, 1976). Possibly alternative energy sources in meat include ribose from *post mortem* break down of ATP and glycerol from lipids (Dainty *et al.*, 1989b). Under aerobic conditions, strains are relatively insensitive to lactic acid which is in contrast to anaerobic conditions (Grau, 1980). This difference is a selective factor in controlling the growth of this organism in vacuum-packaged meat.

4.14.1 Aerobic storage of meat

Gram-negative, non-motile aerobic rods and Gram-positive cocci are frequently isolated from fresh and aerobically stored meat (Barnes *et al.*, 1979; Dainty *et al.*, 1983). Spoilage of these foods is usually attributed to *Pseudomonas* spp. because of their predominance and ability to produce off-odours. Nevertheless, non-motile Gram-negative aerobes can account for 10–15% of the flora after storage reaching numbers of 10^6–10^8/cm^2 which could contribute to spoilage (Shaw and Latty, 1988). A numerical taxonomic study of the non-motile Gram-negative aerobes present in foods resulted in three main clusters grouping around *Acinetobacter johnsonii*, *Psychrobacter immobilis* and non-motile variants of *P. fragi* (Shaw and Latty, 1988).

Spoilage bacteria have been identified as one of the agents responsible for accelerating the oxidation of cherry-red oxymyoglobin to brownish metmyoglobin in fresh meat. When ground beef homogenates were inoculated with fluorescent pseudomonads and/or *Brochothrix thermosphacta* and held at 4°C, an apparent decrease in metmyoglobin content at 10^8/g along with a pH increase occurred (Faustman *et al.*, 1990). During aerobic storage of frozen, thawed muscle total psychrotrophs, *Pseudomonas* spp., *Brochothrix thermosphacta* and Enterobacteriaceae were lowest on PSE pork and highest on DFD pork in comparison to normal pork (Greer and Murray, 1988). Differences in bacterial densities were due to a longer lag phase in PSE pork and a shorter lag phase in DFD pork than in normal pork. Of the three muscle quality groups, DFD was most susceptible to the development of spoilage odours, while PSE pork was most sensitive to deterioration in appearance. *P. fragi*-like organisms play a unique role in producing the combination of esters and sulphur-containing compounds during aerobic chilled storage of meat (Dainty *et al.*, 1989b). Volatile compounds of *P. fragi* and *P. fluorescens* biotype 1 produced during growth on beef at 6°C in air were analysed by Edwards *et al.* (1987a). Compounds of major sensory significance were ethyl and methyl esters of C$_2$–C$_8$ fatty acids and sulphur-containing compounds which included methane and isopropane thiols and their related sulphides and thioesters but not hydrogen sulphide. According to Dainty *et al.* (1989b), volatile compounds associated with microbial growth on beef stored aerobically at chill temperatures are alkyl esters (no methyl esters)

sulphuryl compounds and dimethylsulphide produced by *P. fragi*, hydrogen sulphide produced by *Hafnia alvei, Enterobacter agglomerans, Serratia
liquefaciens, Alteromonas putrefaciens* and *Aeromonas hydrophila*, and
acetoin, diacetyl, 2,3-butanediol and free and esterified branch-chained
alcohols produced by Enterobacteriaceae and *Brochothrix thermosphacta*.
Nicol *et al.* (1970) reported green sulphmyoglobin formation in beef by
hydrogen sulphide producing bacteria. *Pseudomonas mephitica* was identified as the responsible microbe. This organism produced H_2S only when
low oxygen atmospheres were used and the pH of the meat was above 6.0.
Kalchayanand *et al.* (1989) suggested *Clostridium* spp. were responsible for
discoloration of vacuum-packaged beef (spoiled beef was bright pinkish
and changed to greenish colour after 10–12 weeks of storage). The reaction
of hydrogen sulphide with myoglobin to form sulphmyoglobin at low pO_2
and high pH (Nicol *et al.*, 1970) is one possible explanation of the green
surface discolorations on high pH samples. Both the chemical nature of the
green pigment(s) and the mechanism(s) of formation are, however, not
fully understood (Dainty *et al.*, 1989b). Not all hydrogen sulphide-
producing organisms such as *E. agglomerans* can induce green discoloration of meat. On high pH meat, 'greening' was found with *Hafnia alvei,
Serratia liquefaciens, Alteromonas putrefaciens* and *Aeromonas hydrophila*, while 'greening' of normal meat was brought about only by *S.
liquefaciens* (Dainty *et al.*, 1989b). Information about the spoilage potential of other microbes is limited. In the case of the Enterobacteriaceae
strains, which on occasion grow to substantial cell densities, nothing is
known other than their ability to produce hydrogen sulphide. Little is
known of the possible effects of the high pH and the associated low glucose
content of DFD meat on the spoilage properties of any of the common
spoilage organisms (Dainty *et al.*, 1989b).

Despite the undoubted importance of the pseudomonads to the chemical
and sensory changes in naturally contaminated meat, the presence of free
and esterified branched-chain alcohols, and of diacetyl, acetoin, and 2,3-
butanediol shows that minor elements of the flora were also contributing to
the detected chemical changes. Elevated levels of one or more of acetoin,
diacetyl, 2-methylpropanol and 3-methylbutanol may appear at a relatively
early stage of storage of naturally contaminated samples, before either
esters or sulphur-containing compounds are present. Their accumulation
coincides with the detection of creamy, dairy and cheesy odours which
were not sufficiently intense or unpleasant for the meat to be regarded as
spoiled but rather unfresh. Both *Brochothrix thermosphacta* and Enterobacteriaceae must be regarded as potential sources of these compounds.
On high pH meat, Enterobacteriaceae can grow well in excess of $10^7/cm^2$
and metabolise amino acids in the absence of glucose (Dainty *et al.*,
1989b).

4.14.2 Vacuum-packaged meat

The relatively anoxic conditions produced even with films of moderately low oxygen permeability will suppress growth of the strictly aerobic pseudomonads. However, a low meat surface pH is also required to inhibit growth of the facultative anaerobic enterobacteria. During storage at 1°C, vacuum-packaged beef cuts developed spoilage floras of lactobacilli and enterobacteria. Cuts stored under CO_2 developed floras of lactobacilli alone (Gill and Penney, 1986). LAB capable of producing hydrogen sulphide have been isolated from a variety of vacuum-packaged meats (Shay *et al.*, 1988). *Leuconostoc mesenteroides* appeared to be associated with the deterioration in lean colour of vacuum-packaged beef knuckles (Savell *et al.*, 1986). Vacuum-packaged high pH meat will usually be spoiled relatively rapidly by the activities of enterobacteria and associated pH-sensitive facultative anaerobes such as *Alteromonas putrefaciens* and *Brochothrix thermosphacta* (Gill and Penney, 1986). Maximum viable numbers on vacuum-packaged meat are typically below $10^8/cm^2$ and remain at this level for several weeks. The odoriferous end-products of bacterial metabolism progressively accumulate however, yielding the typical off-odour of vacuum-packaged meat. D(−)-lactic acid has been suggested as an indicator of storage time (De Pablo *et al.*, 1989).

An unusual spoilage of chill-stored, normal pH, vacuum-packaged beef by a *Clostridium* spp. (Dainty *et al.*, 1989a) resulted in blowing of the packs and off-odours, perceived on initial opening of the packs as 'sulphurous' and subsequently as 'fruity', 'solvent-like' and 'strong cheese'. Hydrogen and CO_2 were the major components of the headspace gases. Other components included H_2S, methanethiol, dimethyl sulphide, dimethyl di- and trisulphides, methyl-thioacetate, 1-butanol, acetic and butanoic acids, and 1-butyl esters of a range of C_1–C_6 fatty acids. The packs of beef primals had been stored at appropriate temperature for 5–6 weeks. Meat colour was not affected. The total numbers of viable organisms were similar on meat from both the normal and blown commercial packs. LAB were the dominant organisms and *Brochothrix thermosphacta* and Gram-negative organisms were only minor elements of the flora on both types of meat. In experimental inoculated packs the *Clostridium* spp. grew up to $10^8/g$. Surface pH values of meat from the commercial blown packs were higher than those from normal packs (pH 5.6) by 0.3–1.0 pH units. A psychro- trophic *Clostridium* spp. was probably involved in spoilage of vacuum- packaged refrigerated fresh beef stored for 8–10 weeks at 1–3°C (Kalchaya- nand *et al.*, 1989). Accumulation of large quantities of foul-smelling gas was observed along with extensive proteolysis, loss of colour and texture of the meat. Spoilage of vacuum-packaged meat by *Clostridium* spp. at relatively higher storage temperature has been reported by Ingram and Dainty (1971).

4.15 Inhibition of bacteria

LAB produce a variety of antagonistic factors that include metabolic end-products (lactic acid, CO_2, H_2O_2, diacetyl), antibiotic-like substances and bacteriocins (Klaenhammer, 1988). Lactic acid is undoubtedly the most important compound in this respect. Price and Lee (1970) found H_2O_2 inhibitory to *Pseudomonas* spp. and *P. fragi* may be inactivated by hydrogen peroxide produced by *L. bulgaricus* (Gibbs, 1987). *Staphylococcus aureus* was inhibited by peroxide produced by *L. lactis* and *L. bulgaricus* (Dahiya and Speck, 1968). Carbon dioxide inhibits the growth of a wide range of microorganisms (Enfors and Molin, 1980). It may be produced in fermenting sausage in the early ripening stage by aerobic microorganisms until residual oxygen becomes depleted and concomitantly by heterofermentative metabolism of LAB, or by decarboxylation of amino acids.

Lactobacillus spp. are among the least sensitive organisms to high values of pCO_2. *Pseudomonas* spp. are quite sensitive, especially under oxygen limitation. Growth of *E. coli* and *Streptococcus faecalis* is slightly reduced under CO_2, while *Brochothrix thermosphacta* and *Staphylococcus aureus* show lower growth rates. Germination of *Bacillus* spores is inhibited and that of *Clostridium* spores stimulated (Dixon and Kell, 1989). In fermentations, CO_2 pressures may significantly influence the regulation of microbial metabolism favouring biomass or product formation (Dixon and Kell, 1989). The range of inhibitory activity by bacteriocins of LAB can either be narrow, inhibiting only those strains that are closely related to the producer organisms, or wide, inhibiting a diverse group of Gram-positive microorganisms (Klaenhammer, 1988). Bacteriocin of *L. sake* Lb 706 inhibited *L. monocytogenes* in broth, and in minced meat the strain contributed to the inhibition of *L. monocytogenes* (Schillinger and Lücke, 1989, 1990).

4.16 Future developments

Fermented meat aroma is becoming increasingly important in a highly competitive market. This tendency will provoke a variety of new developments in quality control, technology, marketing and starter culture development. Genetic determinants and transfer mechanisms for bacteriocin production are anticipated to contribute significantly to the development and application of genetic technologies to LAB. Strains producing bacteriocins are expected to promote the development and construction of starter cultures for traditional food fermentations which effectively compete with the natural microflora. Microbial contributions to desired lipolytic and proteolytic processes in fermented meats will have to be detailed for individual microorganisms and expressed on a quantitative basis. More quantitative data on key processes in meat fermentations are necessary to improve our understanding of microbial activities in these environments. As an example, the role of oxygen in the fermentation of dry sausages and its effect on the production of H_2O_2 by LAB should be investigated on a more quantitative basis. Similarly, the quantitative contribution of micrococcal catalase to overall catalase activity in fermenting sausages is hardly clear and more recent data are lacking. Differences in microbial abundance within the sausage are difficult to explain by oxygen availability in view of the rapid decrease in pO_2 after stuffing. In general, more data on gas production/consumption of fermenting meats would be of interest. Health and spoilage aspects will continue to stay a relevant field of research, especially when new technologies are applied and new products are created.

References

Abdelal, A.T. (1979) Arginine catabolism by microorganisms. In *Annual Review of Microbiology*, Vol. 33 Eds M.P. Starr, J.L. Ingraham and S. Raffel, pp. 139–168. Annual Reviews, Palo Alto, CA.

Adams, M.R. (1986) Fermented flesh foods. In *Progress in Industrial Microbiology*, Vol. 23 Ed. M.R. Adams, pp. 159–198. Elsevier, Amsterdam.

Alford, J.A., Pierce, D.A. and Suggs, F.G. (1964) Activity of microbial lipases on natural fats and synthetic triglycerides. J. *Lipid Res.*, 5, 390–394.

Anderson, M.E. and Marshall, R.T. (1989) Interaction of concentration and temperature of acetic acid solution on reduction of various species of microorganisms on beef surfaces. J. *Food Protect.*, 52, 312–315.

Andersson, R.E., Daeschel, M.A. and Hassan, H.M. (1988) Antibacterial activity of plantaricin SIK-83, a bacteriocin produced by *Lactobacillus plantarum*. *Biochimie*, 70, 381–390.

Arneth, W. and Herold, B. (1988) Determination of nitrate and nitrite in sausages after enzymatic reduction. *Fleischwirtschaft*, 68, 761–764.

Aryanta, R.W., Fleet G.H. and Buckle K.A. (1991). The occurrence and growth of microorganisms during the fermentation of fish sausage. *Int. J. Food Microbiol.*, 13, 143–156.

Asghar, A., Gray, J.I., Buckley, D.J., Pearson, A.M. and Booren, A.M. (1988) Perspectives on warmed-over flavour. *Food Technol.*, 42, 102–108.

Bacus, J. (1984) *Utilization of Microorganisms in Meat Processing*. Research Studies Press, Letchworth.

Bacus, J.N. and Brown, W.L. (1981) Use of microbial cultures: Meat products. *Food Technol.*, 35, 74–78.

Baird Parker, A.C. and Kilsby, D.C. (1987) Principles of predictive food microbiology. J. *Appl. Bact.* 63, 43S–49S.

Bantleon, A.D. (1987) *Lactobacillus sake* und *Lactobacillus curvatus* als Starterorganismen für die Rohwurstreifung. Thesis, Universität Hohenheim.

Barnes, E.M., Mead, G.C., Impey, C.S. and Adams, B.W. (1979) Spoilage organisms of refrigerated poultry meat. In *Cold Tolerant Microbes in Spoilage and the Environment*. Eds A.D. Russel and R. Fuller, pp. 101–116. Academic Press, London.

Biehl, M.L. and Buck, W.B. (1987) Chemical contaminants: their metabolism and their residues. J. *Food Protect.*, 50, 1058–1073.

Blickstad, E. and Molin, G. (1984) Growth and end-product formation in fermenter cultures of *Brochothrix thermosphacta* ATCC 11509[T] and two psychrotrophic *Lactobacillus* spp. in different gas atmospheres. *J. Appl. Bact.*, **57**, 213–220.

Borch, E. and Molin, G. (1989) The aerobic growth and product formation of *Lactobacillus, Leuconostoc, Brochothrix*, and *Carnobacterium* in batch cultures. *Appl. Microbiol. Biotechnol.*, **30**, 81–88.

Botas, M., Benezet, A., Olmo, N., Osa, J.M. and Perez Florez, F. (1987) Study of phosphorus content in raw cured sausages. *Alimentaria*, **183**, 59–65.

Bottazi, V. (1988) An introduction to rod-shaped lactic-acid bacteria. *Biochimie*, **70**, 303–315.

Bowler, C., Slooten, L., Vandenbranden, S., De Rycke R., Bottermann, J., Sybesma, C., Van Montagu, M. and Inze, D. (1991) Mangan superoxide dismutase can reduce cellular damage mediated by oxygen radicals in transgenic plants. *EMBO J.*, **10**, 1723–1732.

Burrowes, O.J., Schmidt, F.H., Smith, K.L. and Chambers, J.V. (1986) Evaluation of summer sausage manufactured using mixed *Lactobacillus* and *Leuconostoc* starter culture. *J. Food Protect.* **49**, 280–281.

Campanini, M., Mutti, P. and Previdi, P. (1987) Caratterizzazione di Micrococcaceae da insaccati stagionati. *Industria Conserva*, **62**, 3–6.

Campbell-Platt, G. (1987) *Fermented Foods of the World — A Dictionary and Guide*. Butterworths, London.

Cantoni, C. and Bersani, C. (1985). Lactobacilli e maturazione degli insaccati. *Industrie Alimentari*, **24**, 256–258.

Cantoni, C., Molnar, M.R., Renon, P. and Gioletti, G. (1967) Lipolytic micrococci in pork fat. *J. Appl. Bact.*, **30**, 190–196.

Chandler, R.E. and McMeekin, T.A. (1989) Modelling the growth response of *Staphylococcus xylosus* to changes in temperature and glycerol concentration/water activity. *J. Appl. Bact.*, **66**, 543–548.

Collins, E.B. and Aramaki, K. (1980). Production of hydrogen peroxide by *Lactobacillus acidophilus*. *J. Dairy Sci.*, **63**, 353–357.

Collins, F.M. and Carter, P.B. (1978). Growth of Salmonellae in orally infected germfree mice. *Infect. Immun.*, **21**, 41–47.

Collins-Thompson, D.L. and Thomson, I.Q. (1986). Changes in manganese content in *Lactobacillus plantarum* during inhibition with sodium nitrite. *J. Food Protect.*, **49**, 602–604.

Comi, G., Cantoni, C. and Celori, F. (1986) Considerazioni sugli stafilococchi coagulasi negativi degli insaccati crudi stagionati. *Industrie Alimentari*, **25**, 378–380.

Condon, S. (1987) Responses of lactic acid bacteria to oxygen. *FEMS Microbiol. Rev.* **46**, 269–280.

Costerton, J.W., Marrie, T.J. and Cheng, K.-J. (1985) Phenomena of bacterial adhesion. In *Bacterial Adhesion. Mechanisms and Physiological Significance*. eds D.C. Savage and M. Fletcher, pp. 3–43. Plenum Press, New York.

Dahiya, R.S. and Speck, M.L. (1968) Hydrogen peroxide formation by lactobacilli and its effects on *Staphylococcus aureus*. *J. Dairy Sci.*, **51**, 1568–1572.

Dainty, R.H., Shaw, B.G. and Roberts, T.A. (1983) Microbial and chemical changes in chill-stored red meats. In *Food Microbiology: Advances and Prospects*. Eds T.A. Roberts and F.A. Skinner, pp. 151–178. Academic Press, London, New York.

Dainty, R.H., Edwards, R.A. and Hibbards, C.M. (1989a) Spoilage of vacuum-packaged beef by a *Clostridium* sp. *J. Sci. Food Agric.* **49**, 511–516.

Dainty, R.H., Edwards, R.A., Hibbards, C.M. and Ramantanis, S.V. (1986) Bacterial sources of putrescine and cadaverine in chill stored vacuum-packaged beef. *J. Appl. Bact.*, **61**, 117–124.

Dainty, R.H., Edwards, R.A., Hibbards, C.M. and Marnewick, J.J. (1989b) Volatile compounds associated with microbial growth on normal and high pH beef stored at chill temperatures. *J. Appl. Bact.* **66**, 281–290.

Debevere, J.M., Voets, J.P., Schryver, F. and Huyghebaert, A. (1976) Lipolytic activity of *Micrococcus* sp. isolated from a starter culture in pork fat. *Lebensmittel-Wissenschaft Technol.* **9**, 160–162.

Deibel, R.H., Niven, C.F. and Wilson, G.D. (1961) Microbiology of meat curing. III. Some microbiological and related technological aspects in the manufacture of fermented sausages. *Appl. Microbiol.*, **9**, 156–161.

Demasi, T.W., Grimes, L.W., Dick, R.L. and Acton, J.C. (1989) Nitrosoheme pigment formation and light effects on colour properties of semidry nonfermented and fermented sausages. *J. Food Protect.*, **52**, 189–194.

De Pablo, B., Asensio, M.A., Sanz, B. and Ordonez, J.A. (1989) The D(−)lactic acid and acetoin/diacetyl as potential indicators of the microbial quality of vacuum-packaged pork and meat products. *J. Appl. Bact.*, **66**, 185–190.

Dixon, N.M. and Kell, D.B. (1989) The inhibition by CO_2 of the growth and metabolism of microorganisms. *J. Appl. Bact.*, **67**, 109–136.

Dunny, G.M., Krug, D.A., Pan, C.L. and Ledford, R.A. (1988) Identification of cell wall antigens associated with a large conjugative plasmid encoding phage resistance and lactose fermentation ability in lactic streptococci. *Biochimie*, **70**, 443–450.

Edwards, R.A., Dainty, R.H. and Hibbard, C.M. (1987a) Volatile compounds produced by meat pseudomonads and related reference strains during growth on beef stored in air at chill temperatures. *J. Appl. Bact.*, **62**, 403–412.

Edwards, R.A., Dainty, R.H., Hibbard, C.M. and Ramantanis, S.V. (1987b) Amines in fresh beef of normal pH and the role of bacteria in changes in concentration observed during storage in vacuum packs at chill temperatures. *J. Appl. Bact.*, **63**, 427–434.

Egan, A.F. (1983) Lactic acid bacteria of meat products. *Antonie van Leeuwenhoek*, **49**, 327–336.

El-Gendy, S.M., Abdel-Galil, H., Shahin, Y. and Hegazi, F.Z. (1983) Characteristics of salt-tolerant lactic acid bacteria in particular lactobacilli, leuconostocs, and pediococci isolated from salted raw milk. *J. Food Protec.*, **46**, 429–433.

Enfors, S.O. and Molin, G. (1980) Effect of high concentrations of carbon dioxide on growth rate of *Pseudomonas fragi*, *Bacillus cereus* and *Streptococcus cremoris*. *J. Appl. Bact.*, **48**, 409–416.

Farber, J.M. (1986) Predictive modeling of food deterioration and safety. In *Food Borne Microorganisms and their Toxins: Developing Methodology*. eds. M.D. Pierson and N.J. Stern, pp. 57–90. Marcel Dekker, New York.

Faustman, C., Johnson, J.L., Cassens, R.G. and Doyle, M.P. (1990) Colour reversion in beef — Influence of psychrotrophic bacteria. *Fleischwirtschaft int.*, 49–52.

Frey, W. (1987) Schnittfeste Rohwurst im internationalen Vergleich. *Fleischerei*, **38**, 701–703.

Fuller, R. (1989) Probiotics in man and animals. *J. Appl. Bact.*, **66**, 365–378.

Fuller, R. and Brooker, B.E. (1980) The attachment of bacteria to the squamous epithelial cells and its importance in the microecology of the intestine. In *Microbial Adhesion to Surfaces*, eds R.C.W. Berkley, J.M. Lynch, J. Melling, P.R. Rutter and B. Vincent, pp. 495–507. Ellis Horwood, Chichester.

Gardner, G.A. (1981) *Brochothrix thermosphacta (Microbacterium thermosphactum)* in the spoilage of meats: a review. In *Psychrotrophic Microorganisms in Spoilage and Pathogenicity* Eds T.A. Roberts, G. Hobbs, J.H.B. Christian and N. Skovgaard, pp. 139–173. Academic Press, London.

Garriga, M., Compte, M., Casademont, G. and Moreno-Amich, R. (1988) Influence of carbohydrates in the fermentation of dry sausages. *Rev. Agroquim. Tecnol. Alimentos*, **28**, 548–557.

Gehlen, K.H. (1989) Einfluss der Technologie auf die Rohwurstreifung mit *Lactobacillus curvatus, Micrococcus varians* und weiteren Starterorganismen unter besonderer Berücksichtigung der Nitratreduktion. Thesis, Universität Hohenheim.

Giaccone, V., Sibour, M. and Parisi, E. (1988) Sliced vacuum-packaged Italian dry sausages and frankfurter-type sausages. Microbiological processes and shelf-life. *Fleischwirtschaft*, **68**, 1001–1003.

Gibbs, P.A. (1987) Novel uses for lactic acid fermentation in food preservation. *J. Appl. Bact.*, 51S–58S.

Gill, C.O. (1983) Meat spoilage and evaluation of the potential storage life of fresh meat. *J. Food Protect.* **46**, 444–452.

Gill, C.O. and Newton, K.G. (1977) The development of aerobic spoilage flora on meat stored at chill temperature. *J. Appl. Bact.*, **43**, 189–195.

Gill, C.O. and Penney, N. (1986) Packaging conditions for extended storage of chilled dark, firm, dry beef. *Meat Sci.*, **18**, 41–54.

Gilliland, S.E. (1985) *Bacterial Starter Cultures for Foods*. CRC Press, Boca Raton, FL.

Girard, J.P., Bucharles, C., Berdague, J.L. and Ramihone, M. (1989) Einfluss ungesättigter Fette auf Abtrocknungs und Fermentationsvorgänge von Rohwürsten. *Fleischwirtschaft*, **69**, 255–260.

Goldschmidt, M.C. and Lockhart, B.M. (1971) Rapid methods for determining decarboxylase activity: arginine decarboxylase. *Appl. Microbiol.*, **22**, 350–357.

Gonzalez, C.F. and Kunka, B.S. (1987) Plasmid-associated bacteriocin production and sucrose fermentation in *Pediococcus acidilactici*. *Appl. Environ. Microbiol.*, **53**, 2534–2538.

Gottschalk, G. (1979) *Bacterial Metabolism*. Springer-Verlag, New York, Heidelberg, Berlin.

Grau, F.H. (1980) Inhibition of the anaerobic growth of *Brochothrix thermosphacta* by lactic acid. *Appl. Environ. Microbiol.*, **40**, 433–436.

Greer, G.G. and Murray, A.C. (1988) Effects of pork muscle quality on bacterial growth and retail case life. *Meat Sci.*, **24**, 61–72.

Harel, S. and Kanner, J. (1985a) Hydrogen peroxide generation in ground muscle tissues. *J. Agric. Food Chem.*, **33**, 1186–1188.

Harel, S. and Kanner, J. (1985b) Muscle membranal lipid peroxidation initiated by H_2O_2-activated metmyoglobin. *J. Agric. Food Chem.*, **33**, 1188–1192.

Hassan, H.M. and Fridovich, I. (1978) Regulation of the synthesis of catalase and peroxidase in *Escherichia coli. J. Biol. Chem.*, **253**, 6445–6450.

Hastings, J.W. and Holzapfel, W.H. (1987) Conventional taxonomy of lactobacilli surviving radurization of meat. *J. Appl. Bact.*, **62**, 209–216.

Hechelmann, H. and Kasprowiak, R. (1991) Mikrobiologische Kriterien für stabile Produkte. *Fleischwirtschaft*, **71**, 374–389.

Hechelmann, H., Lücke, F.-K. and Schillinger, U. (1988) Ursachen und Vermeidung von *Staphylococcus aureus*-Intoxikationen nach Verzehr von Rohwurst und Rohschinken. *Mitteilungsbl. Bundesanstalt Fleischforsch., Kulmbach*, **100**, 7965–7964.

Hesseltine, C.W. (1984) Fermented Foods in the Orient with emphasis on soy sauce. *J. Japan Soy Sauce Res. Inst.*, **10**, 69–81.

Holley, R.A., Jui, P.A., Wittmann, M. and Kwan, P. (1988) Survival of *S. aureus* and *S. typhimurium* in raw ripened dry sausages formulated with mechanically-separated chicken meat. *Fleischwirtschaft*, **68**, 194–201.

Honkavaara, M. (1988) Influence of PSE pork on the quality and economics of cooked, cured ham and fermented dry sausage manufacture. *Meat Sci.*, **24**, 201–208.

Hugas, M. and Montfort, J.M. (1986) Microbial evolution during the curing of Spanish Serrano hams. The influence of some preservatives on the microbial flora. In *Proc. Eur. Meeting Meat Res. Workers*, **32**/II, 307–310.

Ingram, M. and Dainty, R.H. (1971) Changes caused by microbial spoilage of meat. *J. Appl. Bact.*, **34**, 21–39.

Ishiwa, M. and Iwata, S. (1980) Drug resistance plasmids in *Lactobacillus fermentum. J. Gen. Appl. Microbiol.* **26**, 71–74.

Izumi, K., Cassens, R.G. and Greaser, M.L. (1989) Reaction of nitrite with ascorbic acid and its significant role in nitrite-cured food. *Meat Sci.*, **26**, 141–154.

Jacquet, B. (1988) Les produits du porc. II. Les produits secs. *Viandes et Produits Carnes*, **9**, 67–71.

Johns, A.M., Birkinshaw, L.H. and Ledward, D.A. (1989) Catalysts of lipid oxidation in meat products. *Meat Sci.*, **25**, 209–220.

Joseph, A.L., Berry, B.W., Wagner, S.B. and Davis, L.A. (1978) Lactic acid, pH and bacterial values of dry fermented salami containing mechanically deboned beef and soy protein fiber. *J. Food Protect.*, **41**, 881–884.

Juven, B.J., Weisslowicz, H. and Harel, S. (1988) Detection of hydrogen peroxide produced by meat lactic starter cultures. *J. Appl. Bact.*, **65**, 357–360.

Kagermeier, A. (1981) Taxonomie und Vorkommen von Milchsäurebakterien in Fleischprodukten. Thesis, Ludwig-Maximilian-Universität München.

Kalchayanand, N., Ray, B., Field, R.A. and Johnson, M.C. (1989) Spoilage of vacuum-packaged refrigerated beef by *Clostridium. J. Food Protect.*, **52**, 424–426.

Kandler, O. and Weiss, N. (1986) Genus *Lactobacillus*. In *Bergey's Manual of Systematic Bacteriology*. Eds. P.H.A. Sneath, N.S. Mair, M.E. Sharpe and J.G. Holt. Williams & Wilkins, Baltimore.

Katsaras, K. and Leistner, L. (1988) Topographie der Bakterien in Rohwurst. *Fleischwirtschaft*, **68**, 1295–1298.

Klaenhammer, T.R. (1988) Bacteriocins of lactic acid bacteria. *Biochimie* **70**, 337–349.

Klettner, P.G. and Baumgartner, P.A. (1980) The technology of raw sausage manufacture. *Food Technol. Australia*, **32**, 380–384.

Kok, J. and Venema, G. (1988) Genetics of proteinases of lactic acid bacteria. *Biochimie*, **70**, 475–488.

Kraft, A.A. (1986) Meat Microbiology. In *Muscle as Food*. Ed. P.J. Bechtel, pp. 239–278. Academic Press, Orlando.

Krautil, F.L. and Tulloch, J.D. (1987) Microbiology of mechanically recovered meat. *J. Food Protect.*, **50**, 557–561.

Kroll, R.G., Frears, E.R. and Bayliss, A. (1989) An oxygen electrode-based assay of catalase activity as a rapid method for estimating the bacterial content of foods. *J. Appl. Bact.*, **66**, 209–218.

Lagerborg, V.A. and Clapper, W.E. (1952) Amino acid decarboxylases of lactic acid bacteria. *J. Bact.*, **63**, 393–397.

Landvogt, A. and Fischer, A. (1990) Dry sausage ripening — Targeted control of the acidification capacity of the starter cultures. *Fleischwirtschaft*, **70**, 1134–1140.

Landvogt, A. and Fischer, A. (1991) Rohwurstreifung — Gezielte Steuerung der Säuerungsleistung von Starterkulturen. 2 Teil. *Fleischwirtschaft*, **71**, 32–35.

Law, B.A. and Kolstad, J. (1983). Proteolytic systems in lactic acid bacteria. *Antonie van Leeuwenhoek*, **49**, 225–245.

Leistner, L. (1986) Allgemeines über Rohwurst. *Fleischwirtschaft*, **66**, 290–300.

Leistner, L. (1988) Shelf-stable Oriental meat products. In *Proc. 34th Int. Congr. Meat Sci. Technol., Brisbane, Australia*, pp. 470–475.

Leistner, L. (1991) Fermented and intermediate-moisture meat products. *Outlook on Agriculture*, **20**, 113–119.

Leistner, L. and Lücke, F.-K. (1989) Bioprocessing of meats. In *Biotechnology and Food Processing*. Eds S.D. Kung, D.D. Bills and R. Quatrano, pp. 273–286. Butterworths, Boston.

Leistner, L., Hechelmann, H., Bem, Z. and Albertz, R. (1973) Untersuchungen zur Reduktion des Nitritzusatzes zu Fleischerzeugnissen. *Fleischwirtschaft*, **55**, 1751–1754.

Liepe, H.-U. (1983) Starter cultures in meat production. In *Biotechnology*. Vol. 5. Ed. G. Reed, pp. 400–424. Verlag Chemie, Weinheim.

Liepe, H.U. (1987) Keimverteilung in einer Rohwurst. *Fleischwirtschaft*, **67**, 1266–1267.

Liepe, H.U. (1988) *Lup cheong*. Analysendaten einer chinesischen Trockenwurst. *Fleischwirtschaft*, **68**, 157.

Liepe, H.-U., Pfeil, E. and Porobic, R. (1990) Influence of sugars and bacteria on dry sausage souring. *Fleischwirtschaft int.*, 43–46.

Liu, M.L., Kondo, J.K., Barnes, M.B. and Bartholomew, D.T. (1988) Plasmid-linked maltose utilization in *Lactobacillus* spp. *Biochimie*, **70**, 351–355.

Loewen, P.C. and Switala, J. (1987) Multiple catalases in *Bacillus subtilis*. *J. Bact.*, **169**, 3601–3607.

Lücke, F.K. (1985) Fermented sausages. In *Microbiology of Fermented Foods*, Vol. 2. Ed. B.J.B. Wood, pp. 41–83. Elsevier Applied Science London.

Lücke, F.K., Popp, J. and Kreutzer, R. (1986) Bildung von Wasserstoffperoxid durch Laktobazillen aus Rohwurst und Brühwurstaufschnitt. *Chem. Mikrobiol. Technol. Lebensmittel*, **10**, 78–81.

Marriott, N.G., Naumann, H.D., Stringer, W.C. and Hedrick, H.B. (1967) Colour stability of prepackaged fresh beef as influenced by predisplay environments. *Food Technol.*, **21**, 1518–1520.

Martin, S.E. and Chaven, S. (1987) Synthesis of catalase in *Staphylococcus aureus* MF-31. *Appl. Environ. Microbiol.*, **53**, 1207–1209.

Mattila, T. and Frost, A.J. (1988) The growth of potential food poisoning organisms on chicken and pork muscle surfaces. *J. Appl. Bact.*, **65**, 455–461.

McKay, L. (1986) Application of genetic engineering techniques for dairy starter culture improvement. In *Biotechnology in Food Processing*. Eds S.K. Harlander and T.B. Labuza, pp. 145–156. Noyes Publications, Park Ridge, NJ.

McMeekin, T.A., Chandler, R.E., Doe, P.E., Garland, C.D., Olley, J., Putro, S. and Ratkowsky, D.A. (1987) Model for combined effect of temperature and salt concentration water activity on the growth rate of *Staphylococcus xylosus*. *J. Appl. Bact.*, **62**, 543–550.

Molin, G. (1985) Mixed carbon source utilization of meat-spoilage *Pseudomonas fragi* 72 in relation to oxygen limitation and carbon dioxide inhibition. *Appl. Environ. Microbiol.*, **49**, 1442–1447.

Molin, G. and Ternström, A. (1986) Phenotypically based taxonomy of psychrotrophic *Pseudomonas* isolated from spoiled meat, water and soil. *Int. J. Systematic Bacteriol.*, **36**, 257–274.

Molina, I., Silla, H., Flores, J. and Monzo, J.L. (1990) Study of the microbial flora in dry-cured ham. *Fleischwirtschaft* Int., 54–56.

Montville, T.J., Hsu, A.H.M. and Meyer, M.E. (1987a) High-efficiency conversion of pyruvate to acetoin by *Lactobacillus plantarum* during pH-controlled and fed-batch fermentations. *Appl. Environ. Microbiol.*, **53**, 1798–1802.

Montville, T.J., Meyer, M.E. and Han-Ming Hsu, A. (1987b) Influence of carbon substrates on lactic acid, cell mass and diacetyl–acetoin production in *Lactobacillus plantarum*. *J. Food Protect.*, **50**, 42–46.

Mortvedt, C.I., Nissen-Meyer, J., Sletten, K. and Nes, I.F. (1991) Purification and amino acid sequence of Lactocin S, a bacteriocin produced by *Lactobacillus sake* L45. *Appl. Environ. Microbiol.*, **57**, 1829–1834.

Muriana, P. and Klaenhammer, T.R. (1991a) Purification and partial characterization of Lactacin F, a bacteriocin produced by *Lactobacillus acidophilus* 11088. *Appl. Environ. Microbiol.*, **57**, 114–121.

Muriana, P. & Klaenhammer, T.R. (1991b) Cloning, phenotypic expression, and DNA sequence of the gene for Lactacin F, an antimicrobial peptide produced by *Lactobacillus* spp. *J. Bact.*, **173**, 1779–1788.

Murphy, M.G. and Condon, S. (1984) Comparison of aerobic and anaerobic growth of *Lactobacillus plantarum* in a glucose medium. *Arch. Microbiol.*, **138**, 49–53.

Naes, H., Chrzanowska, J. Nissen-Meyer, J., Pedersen, B.O. and Blom, H. (1991) Fermentation of dry sausage — The importance of proteolytic and lipolytic activities of lactic acid bacteria. *Proc. 37th Int. Congr. Meat Sci. Technol.*, Vol. 2, pp. 914–917.

Nagy, A., Mihalyi, V. and Incze, K. (1988). Lagerung ungarischer Salami. Chemische und organoleptische Veränderungen. *Fleischwirtschaft*, **68**, 431–432, 435.

Newton, K.G. and Gill, C.O. (1978) Storage quality of dark, firm, dry meat. *Appl. Environ. Microbiol.*, **36**, 375–376.

Newton, K.G. and Gill, C.O. (1980) The microbiology of DFD fresh meats: a review. *Meat Sci.*, **5**, 223–232.

Nicol, D.J., Shaw, M.K. and Ledward, D.A. (1970) Hydrogen sulfide production by bacteria and sulfomyoglobin formation in prepackaged chilled beef. *Appl. Microbiol.*, **19**, 937–939.

Niemand, J.G. and Holzapfel, W.H. (1984) Characteristics of lactobacilli isolated from radurized meat. *Int. J. Food Microbiol.*, **1**, 99–110.

Niven, C.F. and Evans, J.B. (1957). *Lactobacillus viridescens* nov. spec., a heterofermentative species that produces a green discolouration of cured meat pigments. *J. Bact.*, **73**, 758–759.

Nurmi, E. (1966) Effect of bacterial inoculations on characteristics and microbial flora of dry sausage. *Acta Agralia Fennica*, **108**, 1–77.

Nychas, G.J., Dillon, V.M. and Board, R.G. (1988) Glucose, the key substrate in the microbiological changes occurring in meat and certain meat products. *Biotechnol. Appl. Biochem.*, **10**, 203–231.

Ordonez, J.A., Asensio, M.A., Garcia, M.L., Selgas, M.D. and Sanz, B. (1990) A reasonable aseptic method of monitoring the phenomena occurring during the ripening of dry fermented sausages. *Fleischwirtschaft Int.*, 54–56.

Papon, M. and Talon, R. (1988) Factors affecting growth and lipase production by meat lactobacilli strains and *Brochothrix thermosphacta*. *J. Appl. Bact.*, **64**, 107–115.

Papon, M. and Talon, R. (1989) Cell location and partial characterization of *Brochothrix thermosphacta* and *Lactobacillus curvatus* lipases. *J. Appl. Bact.*, **66**, 235–242.

Petchsing, U. and Woodburn, M.J. (1990) *Staphylococcus aureus* and *Escherichia coli* in nham (Thai-style fermented pork sausage). *Int. J. Food Microbiol.*, **10**, 183–192.

Poma, J.P. (1987) Prevention du poissage des jambons secs en cours de fabrication. *Viandes et Produits Carnes* **8**, 109–111.

Price, R.J. and Lee, J.S. (1970) Inhibition of *Pseudomonas* species by hydrogen peroxide producing lactobacilli. *J. Milk Food Technol.*, **33**, 13–18.

Puglia, M.L. and Seperich, G.J. (1983) Identification of a stimulatory agent in selected spices. In *83rd Annu. Meeting of the American Society for Microbiology*, Abstract O 83.

Raccach, M. (1981) Control of *Staphylococcus aureus* in dry sausage by newly developed meat starter culture and phenolic-type antioxidants. *J. Food Protect.*, **44**, 665–669.

Raccach, M. (1985) Manganese and lactic acid bacteria. *J. Food Protect.*, **48**, 895–898.

Raccach, M. (1987) Pediococci and biotechnology. *CRC Crit. Rev. Microbiol.*, **14**, 291–309.

Raccach, M. and Baker, R.C. (1978) Formation of hydrogen peroxide by meat starter cultures. *J. Food Protect.*, **41**, 798–799.

Raccach, M., Kovac, S.L. and Mayer, C.M. (1985) Susceptibility of meat lactic acid bacteria to antibiotics. *Food Microbiol.*, **2**, 271–275.

Ramarathnam, N., Rubin, L.J. and Diosady, L.L. (1991) Studies on meat flavour, 1. Qualitative and quantitative at differences in uncured and cured pork. *J. Agric. Food Chem.*, **39**, 344–350.

Ray, B., Johnson, C. and Field, R.A. (1984) Growth of indicator, pathogenic and psychrotrophic bacteria in mechanically separated beef, lean ground beef and beef bone marrow. *J. Food Protect.*, **47**, 672–677.

Reuter, G. (1970) Laktobazillen und eng verwandte Mikroorganismen in Fleisch und Fleischerzeugnissen. *Fleischwirtschaft*, **50**, 954–962.

Rhee, K.S. (1988) Enzymic and nonenzymic catalysis of lipid oxidation in muscle foods. *Food Technol.*, **42**, 127–132.

Rheinbaben, K.E. and Seipp, H. (1986) Studies on the microflora of raw hams with special reference to Micrococcaceae. *Chemie Mikrobiol. Technol. Lebensmittel*, **9**, 152–161.

Rice, S.L., Eitenmiller, R.R. and Koehler, P.E. (1975) Histamine and tyramine content of meat products. *J. Milk Food Technol.*, **38**, 256–258.

Rieman, H., Lee W.H. and Genigeorgis, C. (1972). Control of *Clostridium botulinum* and *Staphylococcus aureus* in semi-preserved meat products. *J. Milk Food Technol.*, **35**, 514–523.

Roberts, T.A. and Jarvis, B. (1983) Predictive modelling of food safety with particular reference to *Clostridium botulinum* in model cured meat systems. In *Food Microbiology Advances and Prospects* eds T.A. Roberts and B. Jarvis, pp. 85–95. Academic Press, London.

Roca, M. and Kalman, I. (1989) Antagonistic effect of some starter cultures on Enterobacteriaceae (*E. coli*). *Meat Sci.*, **25**, 123–132.

Rödel, W., Stiebing, A. and Kröckel, L. (1991) Traditionelle Rohwurst mit Schimmelbelag. Entwicklung eines Herstellungsstandards für Salami. *Bundesministerium für Landwirtschaft, Ernährung und Forsten, Forschungsreport* **6**, 18–19.

Rödel, W., Scheuer, R., Stiebing, A. and Klettner, P.G. (1990). Messung des Sauerstoffgehaltes in Fleischerzeugnissen. *Mitteilungsb. Bundesanstalt Fleischforsch. Kulmbach*, **29**, 53–60.

Santos-Buelga, C., Pena-Egido, M.J. and Rivas-Gonzalo, J.C. (1986). Changes in tyramine during *chorizo*-sausage ripening. *J. Food Sci.*, **51**, 518–519, 527.

Sanz, B., Selgas, D., Parejo, I. and Ordonez, J.A. (1988) Characteristics of meat lactobacilli isolated from dry fermented sausages. *Int. J. Food Microbiol.*, **6**, 199–205.

Savell, J.W., Griffin, D.B., Dill, C.W., Acuff, G.R. and Vanderzant, C. (1986) A research note: Effect of film oxygen transmission rate on lean colour and microbiological characteristics of vacuum-packaged beef knuckles. *J. Food Protect.*, **49**, 917–919.

Savic, Z., Zhang, K.S. and Savic, I. (1988) Chinese-style sausages. A special class of meat products. *Fleischwirtschaft*, **68**, 612–617.

Scharner, E. (1990) Verdrängungseffekte von Starterkulturen gegenüber Salmonellen in Rohwürsten. *Fleischwirtschaft*, **70**, 1183–1186.

Schiefer, G. and Schöne, R. (1978) Rohwurstherstellung mittels Starterkulturen. *Nahrung*, **22**, 419–424.

Schillinger, U. and Lücke, F.-K. (1987) Identification of lactobacilli from meat and meat products. *Food Microbiol.*, **4**, 199–208.

Schillinger, U. and Lücke, F.K. (1989) Antibacterial activity of *Lactobacillus sake* isolated from meat. *Appl. Environ. Microbiol.*, **55**, 1901–1906.

Schillinger, U. and Lücke, F.K. (1990) Lactic acid bacteria as protective cultures in meat products. *Fleischwirtschaft*, **70**, 1296–1299.

Schleifer, K.H. (1986) Section 12. Gram-positive cocci. In *Bergey's Manual of Systematic Bacteriology*, Vol. 2. Eds. P.H.A. Sneath, N.S. Mair, M.E. Sharpe and J.G. Holt. Williams & Wilkins, Baltimore, London, Los Angeles, Sydney.

Seager, M.S., Banks, J.G., Blackburu, W. and Board, R.G. (1986) A taxonomic study of *Staphylococcus* spp. isolated from fermented sausages. *J. Food Sci.*, **51**, 295–297.

Sedewitz, B., Schleifer, K.H. & Götz, F. (1984) Physiological role of pyruvate oxidase in the aerobic metabolism of *Lactobacillus plantarum*. *J. Bact.*, **160**, 462–465.

Selgas, M.D., Sanz, B. and Ordonez, J.A. (1988). Selected characteristics of micrococci isolated from Spanish dry fermented sausages. *Food Microbiol.*, **5**, 185–193.

Shahidi, F., Rubin, L.J. and Wood, D.F. (1988). Stabilization of meat lipids with nitrite-free curing mixtures. *Meat Sci.*, **22**, 73–80.

Shaw, B.G. and Harding, C.D. (1984). A numerical taxonomic study of lactic acid bacteria from vacuum-packaged beef, pork, lamb and bacon. *J. Appl. Bact.*, **56**, 25–40.

Shaw, B.G. and Latty, J.B. (1988) A numerical taxonomic study of non-motile non-fermentative Gram-negative bacteria from foods. *J. Appl. Bact.*, **65**, 7–22.

Shay, B.J., Egan, A.F., Wright, M. and Rogers, P.J. (1988) Cysteine metabolism in an isolate of *Lactobacillus sake*: plasmid composition and cysteine transport. *FEMS Lett.*, **56**, 183–188.

Silla, H., Molina, I., Flores, J. and D. Silvestre (1989) Study of the microbial flora in dry-cured ham. 1. Isolation and growth. *Fleischwirtschaft*, **69**, 1128–1131.

Slemr, J. (1981) Biogene Amine als potentieller chemischer Qualitätsindikator für Fleisch. *Fleischwirtschaft*, **61**, 921–926.

Sneath, P.H.A. and Jones, D. (1976) *Brochothrix*: a new genus tentatively placed in the family Lactobacillaceae. *Int. J. Systematic Bact.*, **26**, 102–104.

Stackebrandt, E. and Teuber, M. (1988) Molecular taxonomy and phylogenetic position of lactic acid bacteria. *Biochimie*, **70**, 317–324.

Stengel, G. (1990) Staphylococci. *Fleischwirtschaft Int.*, 62–66.

Stiebing, A. and Rödel, W. (1988) Influence of relative humidity on the ripening of dry sausage. *Fleischwirtschaft*, **68**, 1287–1291.

Tetlow, A.L. and Hoover, D.G. (1988) A research note: Fermentation products from carbohydrate metabolism in *Pediococcus pentosaceus* PC39. *J. Food Protect.* **51**, 804–806.

Thompson, J. (1988) Lactic acid bacteria: model systems for *in vivo* studies of sugar transport and metabolism in Gram-positive organisms. *Biochimie* **70**, 325–336.

Vandendriessche, F., Vandekerckhove, P. and Demeyer, D. (1980) The influence of some spices on the fermentation of a Belgian dry sausage. In *Proc. 26th Eur. Meeting Meat Res. Workers, Colorado Springs*, Vol. 2, pp. 128–133.

Vidal, C.A. and Collins-Thompson, D.L. (1987) Resistance and sensitivity of meat lactic acid bacteria to antibiotics. *J. Food Protect.*, **50**, 737–740.

von Husby, K.O. and Nes, I.F. (1986) Changes in the plasmid profile of *Lactobacillus plantarum* obtained from commercial meat starter cultures. *J. Appl. Bact.*, **60**, 413–418.

Wadström, T., Andersson, K., Sydow, M., Axelsson, L., Lindgren, S. and Gullmar, B. (1987) Surface properties of lactobacilli isolated from the small intestine of pigs. *J. Appl. Bact.*, **62**, 513–520.

Weidenfeller, P. and Fegeler, W. (1990). Methodological aspects of a micro-identification technique for the differentiation of coagulase-negative staphylococci to species level. *Zentralb. Bakteriol.*, **274**, 78–90.

Wiegner, J. and Hildebrandt, G. (1986) Zur Mindesthaltbarkeit von vakuumverpacktem Brühwurstaufschnitt. *Fleischwirtschaft*, **66**, 316–322.

Winter, R. (1988) Zuckerstoffe und GDL für die Rohwurstherstellung: Einflüsse auf die Reifung. *Fleischerei*, **39**, 843–844.

Wirth, F. (1986) The technology of processing meat not of standard quality. *Fleischwirtschaft*, **66**, 1256–1260.

Wood, B.J.B. (1985) *Microbiology of Fermented Foods*, Vols. 1 and 2. Elsevier Applied Science, London.

Zee, J.A., Simard, R.E., Vaillancourt, R. and Boudreau, A. (1981). Effect of *Lactobacillus brevis, Saccharomyces uvarum* and grist composition on amine formations in beers. *Can. Inst. Food Sci. Technol. J.*, **14**, 321–325.

5 Fungal ripened meats and meat products

P.E. COOK

5.1 Introduction

Fungi have an important role in the production of fermented foods, particularly in the Orient where they are involved in the production of a number of savoury foods, condiments, sweet desserts and alcoholic beverages (Leistner, 1986; Campbell-Platt, 1987; Campbell-Platt and Cook, 1989; Cook and Campbell-Platt, 1994). Although the use of moulds for food production is less extensive in Europe and North America, they still have a significant role in the production of Brie, Camembert and blue vein cheeses as well as some types of raw fermented sausage and ham.

Moulds have probably been used in meat fermentation and ripening for many centuries. Sausage fermentation is thought to be some 250 years old and the dry curing of hams and other meats is likely to be based on much older technology. As with many cheese fermentations, the techniques developed for sausage and ham preservation were developed locally as 'cottage industries' using a diverse range of processes and resulting in products with a variable microbial flora. Many fermented meat products are characterised by the presence of a surface mould flora which, in some cases, may entirely cover the surface of the meat. Local preferences may influence whether fungi are allowed to grow on the surface of the meat. Whereas a surface mould growth is intentionally cultivated or encouraged in parts of Hungary, Romania, Bulgaria, France, Switzerland, Austria, Italy and Spain, in other regions it may be positively avoided or removed (Leistner, 1986).

5.2 The fungal flora of non-fermented meats

5.2.1 Moulds

Although a large number of mould species are associated with plant foodstuffs, the mould flora of meat is more limited and develops following post-slaughter contamination from the air, water, soil processing facilities or by the addition of food ingredients. Moulds are rarely associated with the spoilage of fresh meats largely because they are outgrown by bacteria which can proliferate more rapidly and induce organoleptic changes in the meat. The mould flora of fresh meat does not differ significantly from that associated other foodstuffs, the major genera isolated from fresh meats and meat products being *Alternaria*, *Aspergillus*, *Botrytis*, *Cladosporium*, *Fusarium*, *Geotrichum*, *Monascus*, *Mucor*, *Neurospora*, *Penicillium*, *Rhizopus* and *Thamnidium* (Jay, 1978; Moreau, 1979).

Williams (1990) reported a list of spoiled food items in the UK from which species of *Aspergillus* and *Penicillium* had been isolated. Moulds isolated from spoiled raw meat or sausages included *Penicillium aurantiogriseum*, *P. chrysogenum*, *P. commune*, *P. crustosum*, *P. cyclopium* and *P. expansum*. Aspergilli were not isolated from these products. Frisvad (1988) also included these penicillia in his list of important species from meat but also included *P. brevicompactum*, *P. nalgiovense*, *P. verrucosum*, *P. glabrum*, *P. variabile* and *P. roquefortii*. He also listed *Aspergillus versicolor*, *A. niger*, *A. flavus*, *A. restrictus* and *Eurotium* spp. as being important species associated with meat. Table 5.1 lists a selection of mould species which have frequently been isolated from meats.

5.2.2 Yeasts

Like moulds, yeasts are usually present in low numbers on fresh meat but can compete with bacteria if the surface of the meat becomes dry or competition with bacteria is reduced due to the presence of sulphite (Dillon and Board, 1991). Jay and Margitic (1981) reported yeast counts of 200 to 6.2×10^4/g on fresh ground beef compared with counts of Gram-negative bacteria of 2×10^3 to 6.6×10^7/g. During low-temperature storage of meat, yeast counts may increase and eventually dominate the microflora. Lowry and Gill (1984) observed that yeast counts on the loins of lamb packaged in gas-permeable plastic film increased from 10/cm^2 to 10^6/cm^2 after storage for 20 weeks at $-5°C$ suggesting successful competition with the psychrotrophic bacteria flora. In a review by Jay (1978), species of *Candida*, *Debaryomyces* and *Torulopsis* were listed as the most frequently isolated genera from meats. Other genera associated with fresh meat include *Bullera*, *Cryptococcus*, *Pichia*, *Saccharomyces*, *Schizosaccharomyces*, *Torulaspora*, *Trichosporon* and *Williopsis*.

On processed meats such as sausages, burgers, luncheon meat and smoked ham *Debaryomyces hansenii*, *Trichosporon* spp., and *Candida* spp. such as *Candida zeylanoides* may form a significant component (Jay, 1978; McCarthy and Damoglou, 1993; Viljoen *et al.*, 1993). Although numbers of yeasts on meat are generally lower than the numbers of spoilage bacteria they can occasionally proliferate to high numbers forming a visible surface slime, particularly on some types of sausage (Cesari and Guilliermond,

Table 5.1 Yeast and moulds frequently isolated from raw and fermented meats*

Yeasts	Moulds
Candida catenulata	Alternaria tenuis
Candida curvata	Aspergillus niger
Candida humicola	Aspergillus restrictus
Candida lipolytica	Aspergillus versicolor
Candida mesenterica	Aureobasidium pullulans
Candida parapsilosis	Cladosporium cladosporioides
Candida rugosa	Cladosporium herbarum
Candida sake	Eurotium amstelodami
Candida versatilis	Eurotium chevalieri
Candida vini	Eurotium halophilicum
Candida zeylanoides	Eurotium repens
Cryptococcus albidus	Eurotium rubrum
Cryptococcus curvatus	Geomyces pannorum
Cryptococcus infirmominiatus	Geotrichum candidum
Cryptococcus laurentii	Penicillium aurantiogriseum
Debaryomyces hansenii	Penicillium commune
Debaryomyces kloeckeri	Penicillium chrysogenum
Debaryomyces nicotaine	Penicillium corylophilum
Debaryomyces subglobosus	Penicillium crustosum
Hansenula spp.	Penicillium expansum
Hypopichia burtonii	Penicillium frequentens
Leucosporidium scotti	Penicillium glabrum
Pichia carsonii	Penicillium janthinellum
Pichia fermentans	Penicillium nalgiovense
Pichia guilliermondii	Penicillium notatum
Pichia membranaefaciens	Penicillium variable
Rhodosporidium infirmo-miniatum	Penicillium verrucosum var. cyclopium
Rhodotorula glutinis	Rhizopus nigricans
Rhodotorula mucilaginosa	Scopulariopsis spp.
Rhodotorula rubra	Thamnidium elegans
Torulopsis candida	
Trichosporon cutaneum	
Trichosporon pullulans	
Yarrowia lipolytica	

* Information from: Leistner and Ayres (1968); Hadlok et al. (1975, 1976); Smith and Hadlock (1976); Jay (1978); Dillon and Board (1991); Williams (1990); McCarthy and Damoglou (1993).

1920; Mrak and Bonar, 1938; Drake et al., 1958). Yeasts commonly isolated from meats are listed in Table 5.1.

5.2.3 Cold-stored and frozen meat

Although the spoilage of fresh meat is mostly associated with bacteria, fungi, and in particular moulds, are capable of growing and causing spoilage of meat during chill or frozen storage (Semeniuk and Bell, 1937). Schmidt-Lorenz and Gutschmidt (1969) found that moulds and yeasts could grow on poultry during storage for one year at −10°C and Michener and Elliot (1964) cited references where fungal growth was reported to occur at temperatures as low as −12°C. Brooks and Hansford (1923) were able to isolate *Sporotrichum carnis* (= *Geomyces pannorum*), *Cladosporium herbarum*, *Mucor mucedo*, *Mucor racemosus*, *Penicillium glaucum*, *Thamnidium elegans* and *Helicostylum pulchrum* (= *Chaetostylum fresenii*) from frozen meat. Reports in the literature during the 1920s mainly attributed low-temperature spoilage of meat to either mucoraceous fungi which grow on meat as so-called 'whiskers' or *Cladosporium* spp. growing on meat and forming 'black spots' (Brooks and Kidd, 1921; Brooks and Hansford, 1923; Haines, 1931). With changes in meat storage practices over the last 50 years, spoilage of meat, particularly by mucoraceous fungi, appears to be less common although the intentional growing of mucoraceous fungi on meat has been considered as a means of achieving accelerated ageing (Williams, 1957, 1962; Anon, 1978).

Although the black spots which develop on meat at low temperature were originally thought to be due to *Cladosporium* spp., Gill *et al.* (1982) demonstrated that a number of species of mould may be involved since they were able to culture *Aureobasidium pullulans*, *Cladosporium cladosporioides*, *Cladosporium herbarum* and *Penicillium hirsutum*. They further demonstrated that these fungi were able to grow and produce 'black spots' on meat at −1°C and were even capable of growing at temperatures as low as −5°C. However, black spot spoilage was considered to occur primarily at higher temperatures when the faster-growing psychrotrophic spoilage bacteria were limited by surface desiccation rather than the temperature (Gill and Lowry, 1982). Lowry and Gill (1984) examined the temperature and water activity limits of a number of moulds isolated from meat. A temperature of −5°C was considered to be the practical limit for mould growth on meat after taking into account the additional effects on growth by a reduced water activity.

Cladosporium spp. have also been isolated from fermented sausages and country-cured hams where they are associated with black spot defects similar to those on chill stored or frozen meat. Leistner and Ayres (1968) and Huerta *et al.* (1987a) reported *Cladosporium* spp. from 30% of raw ham samples and Skrinjar and Horvat-Skenderovic (1989) isolated *Cladosporium* from 59% of 31 samples of tea sausage (Teewurst) in the former Czechoslovakia. Although black spots on fermented meats are thought to be due to *Cladosporium* spp., no detailed studies appear to have been made of their composition. In view of the diverse microflora associated with fermented meats, the black spots on these products are unlikely to be due solely to *Cladosporium* spp.

Many yeasts are capable of growing on meat stored at low temperatures (Scott, 1936; Dykes *et al.*, 1991) and Scott (1936) reported that off-flavours from chilled beef were due to lipolytic activity by yeasts. Lowry and Gill (1984) found that the yeast flora of lamb loins increased by five log cycles after storage for five months in an oxygen-permeable plastic film at −5°C.

The yeast flora was composed of *Cryptococcus laurentii* var. *laurentii, Crypt. infirmominiatus, Trichosporon pullulans* and *Crypt. zeylanoides* with *Crypt. laurentii* var. *laurentii* comprising 90% of the yeast population.

5.3 Accelerated ageing of meat using fungi

Before meat is distributed to the consumer, it is often stored under conditions of controlled temperature and relative humidity. In the normal ageing process, meat such as beef may be hung in refrigerated storage to enable tenderisation by proteolytic enzymes in the meat. Low temperatures are required to avoid bacterial spoilage and meat tenderisation by endogenous enzymes may require periods in excess of three weeks.

During this period of ageing, mucoraceous moulds of the genera *Mucor, Rhizopus* and *Thamnidium* may contaminate and grow on the meat forming patches frequently described as 'whiskers'. These surface growths of mucoraceous moulds have been implicated in the tenderisation of meat such as beef and attempts have been made to exploit the proteolytic ability of these fungi by intentional inoculation (Anon, 1978).

Williams (1957) patented a method for ageing meat by intentionally inoculating the surface of beef with the mould *Thamnidium elegans*. The meat was aged at between 7–16°C at relative humidities greater than 80%. Application of *Thamnidium elegans* spores resulted in the meat being tenderised in 3–4 days without undesirable growth of bacteria. In a later patent, Williams (1962) proposed a method which domestic consumers could use to tenderise meat. One approach was to incorporate spores of *Thamnidium elegans* and an organic acid such as citric acid powder into a cellulose sponge or sawdust shavings encased in a permeable nylon cover. This was placed in close proximity to the meat to allow spores or mycelial growth to transfer and the enzymes to tenderise the meat. Incubation of beef steak in close proximity to the inoculum for 48 hours at 4.4°C was found to be satisfactory for tenderising the tissues and producing an acceptable flavour.

Mucoraceous moulds are well known for their proteolytic ability and as sources of milk clotting enzymes for the dairy industry (Ellis *et al.*, 1974; Campbell-Platt and Cook, 1989). *Thamnidium elegans* is thought to release proteolytic enzymes which tenderise beef (Ingram and Dainty, 1971) and it has also been reported to produce lipolytic enzymes (Alford *et al.*, 1964). Kotula *et al.* (1982) examined the proteolytic and lipolytic ability of several moulds which grow on aged beef. There was negligible proteolytic and lipolytic activity at 4°C unless a long incubation period was used suggesting that any fungal tenderisation of beef at low temperatures is likely to be minimal. At higher temperatures (18 and 24°C), there was more pronounced proteolytic and lipolytic activity but inoculation of beef with fungi at these temperatures would be impractical since bacterial growth would also be favoured.

5.4 Fungi flora of fermented meats

5.4.1 House flora

Many of the traditional fermentations which involve fungal ripening rely on passive inoculation by the indigenous 'house flora' which are associated with the building, rooms, caves and equipment used for fermentation, storage and maturation. Examples of passive inoculation by a mould 'house flora' occur in other food fermentations such as cheese fermentations involving *Penicillium camembertii* and *Penicillium roquefortii* and koji production for soy sauce using *Aspergillus* spp. (Leistner, 1986; Cook and Campbell-Platt, 1994). Some producers still rely on the 'house flora' to provide a natural inoculum for mould coating and ripening of salami and raw ham although with concerns about safety, the use of defined fungal starter cultures is likely to become more widespread even for the production of fermented hams. The spontaneous and heterogeneous nature of the house mycoflora can lead to faulty products (Spotti *et al.*, 1988; Le Bars and Le Bars, 1993) and, in particular, a risk of mycotoxins being produced in the product (see Chapter 9). Since many of the fungi isolated from ripened hams and salami are known mycotoxin producers, there has been increasing concern about the continued use of undefined strains of moulds for meat fermentation (Leistner *et al.*, 1989). Some producers actively discourage mould growth on sausages and hams by dipping the meat in 5–10% (w/v) potassium sorbate or 2000 ppm pimaricin (Kemp *et al.*, 1983; Holley, 1986).

As with many other fungal fermented foods, the house flora consists of a wide range of different fungi and the colonisation of the meat is influenced by environmental (temperature, relative humidity, spore concentrations in the atmosphere) as well as factors relating to the meat product (salt concentration, fat content, casing type, water activity). Leistner and Ayres (1968) conducted an extensive survey of the mould flora of fermented meats including salami-type sausage and country-cured hams. *Penicillium* spp. together with *Aspergillus* and *Eurotium* spp. formed a significant proportion of the mould flora, particularly in the case of country-cured hams. The mycoflora of mould-fermented sausages tend to be dominated by *Penicillium* spp., but dry-cured hams are aged for a longer period (up to two years) and develop a different mycoflora perhaps as a result of attaining a lower water activity near the surface of the ham. *Aspergillus* and *Eurotium* spp. develop more extensively on country-cured hams and the presence of particular species may provide some indication of the degree of ripening (Leistner and Ayres, 1968; Monte *et al.*, 1986; Huerta *et al.*, 1987*a*). In sausage production the presence of *Aspergillus* and *Eurotium* spp. is regarded as undesirable since *Penicillium* spp. are considered to be the most important moulds during ripening.

Andersen (1993) conducted an extensive survey of the mycoflora of mould-fermented sausages in Europe and found that *Penicillium* constituted 96% of the mycoflora with species of *Aspergillus*, *Cladosporium*, *Eurotium*, *Mucor*, *Wallemia* and yeasts forming only a minor component. *Penicillium nalgiovense* formed 50% of the mycoflora with *P. chrysogenum*, *P. verrucosum*, *P. oxalicum* and *P. commune* comprising 10, 5, 3 and 3% of the mycoflora, respectively. Fielder (1973) obtained 123 mould isolates from 63 moulded meat products. *Penicillium* spp. comprised 63% of the isolates and *Aspergillus* spp. 24%.

In Italy, Dragoni *et al.* (1980) isolated *Aspergillus candidus*, *A. flavus*, *A. fumigatus*, *A. caespitosus*, *A. niger*, *A. sulfureus*, *A. wentii*, *Penicillium verrucosum* var. *cyclopium* and *Rhizopus nigricans* from the surface of 40 Parma and San Daniele raw ripened hams. Some 90% of the aspergilli isolated were of potentially toxinogenic strains although toxin production was not demonstrated. Huerta *et al.* (1987a) and Rojas *et al.* (1991) also found that *Aspergillus* and *Eurotium* species dominated the fungal flora of Spanish dry-cured hams with *Penicillium* spp. being less frequent on the hams as maturation progressed. Other genera were isolated in these studies (*Alternaria*, *Aureobasidium*, *Cladosporium*, *Fusarium*, *Monilia*, *Paecilomyces*), but their incidence was always much lower than for *Aspergillus* and *Penicillium*.

Table 5.2 shows the frequency of mould genera isolated in a survey of country-cured hams in the USA and dry-cured sausages in Europe (Ayres *et al.*, 1974). Although *Penicillium* spp. are the most important in terms of frequency, more genera were isolated from the hams and it is noticeable that the frequency of *Aspergillus* was much higher on hams than on dry-cured sausage.

5.4.2 Yeasts

The yeast flora of fermented raw sausages and hams has been less intensively studied than the mould flora and most studies have focused on fermented sausages rather than hams (Deak and Beuchat, 1987). Leistner and Bem (1970) and Monte *et al.* (1986) found *Debaromyces hansenii* to be the most frequent yeast on fermented meats although *Candida rugosa*, *Candida catenulata* and *Yarrowia lipolytica* were also isolated. Smith and Hadlock (1976) isolated species of *Candida*, *Cryptococcus*, *Debaryomyces*, *Pichia*, *Rhodotorula* and *Trichosporon*.

Table 5.2 Frequency of mould genera isolated from 400 country-cured hams in the USA and from 40 salamis in Europe. (Data modified from Ayres et al., 1974)

Genus	Frequency (%)*	
	Ham	Salami
Penicillium	63.1	51.8
Aspergillus**	29.6	14.0
(Eurotium)	(15.5)	(12.1)
(Emericella)	(0.7)	not detected
Scopulariopsis	0.3	18.8
Rhizopus	0.7	3.5
Fusarium	0.7	not detected
Epicoccum	0.4	not detected
Cladosporium	2.6	not detected
Alternaria	2.0	not detected
Geotrichum	0.1	4.7
Paecilomyces	not detected	3.5
Mucor	0.4	2.4
Mortierella	0.1	1.2
Syncephalastrum	0.1	not detected

* Based on 758 isolates from country-cured hams and 85 isolates from salami. There were 980 original mould isolates of which 137 were unidentified.
** Frequency includes teleomorph isolates.

Yeast counts in fermented meats are usually low in comparison with those of the bacteria suggesting that any role in the ripening stages of fermentation is likely to be minimal. Huerta et al. (1988) studied the yeast flora of dry-cured Spanish hams at different stages of processing. Numbers were 10^3/g after the addition of salt rising to 10^6/g in the middle of the fermentation and falling to 10^4/g at the end of the curing period. The yeasts were isolated from a layer 0.2–0.4 cm from the surface, suggesting that extensive penetration of the tissues was minimal. The dominant yeasts throughout the fermentation were *Debaryomyces* spp. Species of *Rhodotorula* and *Cryptococcus* were found in the initial stages of fermentation and *Candida* and *Saccharomycopsis* species occurred in low numbers. All yeasts grew in 8% NaCl and most were lipolytic but not proteolytic. Besançon et al. (1992) found *Debaryomyces hansenii* and its non-sporulating form *Candida famata* to be the dominant yeast on the surface of Roquefort cheese. A common factor between fermented meats and some cheeses like Roquefort is the high salt content. Besançon et al. (1992) demonstrated that some strains of *Debaromyces hansenii* could tolerate up to 20% NaCl.

Molina et al. (1990) in a study of the yeasts of slow and fast dry-cured ham isolated species of *Hansenula*, *Cryptococcus*, *Rhodotorula* and *Debaromyces*. Of 102 isolates, 67% were *Hansenula*, 19% *Rhodotorula*, 9% *Cryptococcus* and 5% *Debaryomyces*. *Hansenula citerrii* and *Hansenula holstii* were the most frequent Hansenula species.

5.5 Fermentation

Sausages are made of a mixture of chopped meat, fat, spices and salts which together form coagulated structure inside of a casing. The addition of sodium chloride increases the water binding capacity and leads to a loss of the typical banding pattern in the muscle tissue (Katsaras and Budras, 1992). Further structural changes occur through gradual drying of the meat combined with the accumulation of organic acids from bacterial fermentation. However, little is known of the influence of mould ripening on sausage structure. Although visible mould growth occurs on the outer casing of sausages, Grazia et al. (1986) observed that in inoculation experiments a range of moulds were able to grow throughout the sausage including the centre. This may account for the decrease in lactic acid and rise in pH which has been observed by some researchers (Grazia et al., 1986; Roncales et al., 1991).

The water activity (a_w) of fermented meats varies depending on the size of the meat, length of ripening, salt and fat content, casing permeability, temperature and the relative humidity of the air. Leistner and Wirth (1972) reported that the range in a_w for raw fermented sausage was between 0.83 and 0.96 with a mean of 0.91. The a_w of raw ham varied between 0.86 and 0.93 with a mean of 0.93 and for specific types of meats the a_w values were Teewurst 0.95, cervelat 0.91, salami 0.90 and Hungarian salami 0.83. Non-fermented meats such as frankfurters, luncheon meat and liver sausage ranged between 0.95 and 0.98. Leon et al. (1988) measured the a_w of Serrano hams at different stages of production and found that the a_w varied between 0.83 and 0.97. Rojas et al. (1991) reported a_w values of 0.84–0.95 for Spanish dry-cured hams with the a_w at the end of maturation being about 0.90. Italian and American hams have an a_w of between 0.91 and 0.94 (Dalla Rosa et al., 1987; Huerta et al., 1987a). However, although these figures are representative of fermented meats as a whole, they do not necessarily reflect the water activities close to the surface of these products. In these situations, the a_w may be somewhat lower since xerophilic moulds such as Eurotium spp. grow on the surface of dry-cured hams particularly towards the end of maturation (Huerta et al., 1987a).

5.5.1 Changes in numbers of fungi

Most studies report the frequencies of different fungi from fermented meats and few studies have examined changes in numbers presumably during ripening presumably because of the difficulty in relating numbers of colony forming units to mycelial growth and biomass. Viable counts are likely to be a reflection of spore production rather than mycelial growth since hyphae are likely to be firmly bound to the substrate.

Roncalés et al. (1991) examined mould and yeast growth on fermented sausages over a 28-day period. The counts of moulds and yeasts were initially 10^2–10^3/cm^2, they increased markedly in the first 10 days of ripening and after 25 days numbers had risen to 10^6–10^7/cm^2 and 10^5–10^6/cm^2 for moulds and yeasts, respectively. Most mould growth is restricted to the outer surfaces of fermented meats although Grazia *et al.*, (1986) observed that moulds inoculated onto salami could be isolated from the innermost parts of the product.

5.5.2 Organic acids

During sausage fermentation, the pH value decreases rapidly from an initial pH 5.9–6.2 to around 5.2–5.4 some 5–10 days after casing. In those meat products which involve mould ripening, the pH often rises again, presumably due to the utilisation of lactate and acetate and/or the production of ammonia from protein hydrolysis. Roncalés *et al.* (1991) examined the effect of natural pork casing and artificial collagen casing on the ripening and sensory quality of mould-covered dry sausage. Mould colonisation was more extensive on the natural pork casing and appeared to penetrate and disintegrate the tissues. The pH value of sausage with a natural casing was significantly higher after 10 days of ripening (around 5.6) than sausage with a fibrous collagen casing (around 5.2–5.3). Grazia *et al.* (1986) examined selected physical and chemical characteristics of salami inoculated with nine *Penicillium* strains and one *Aspergillus* strain isolated from sausage factories. After 30 days' fermentation with selected moulds, the pH values were between 6.02 and 6.50 whereas control salami which had been treated with sorbate to prevent mould growth had a pH value of 5.34. Ammonia was detected in all samples but in only four of the mould cultures was it higher than the control treated with sorbate. The pH changes observed in mould-ripened sausages are well within the growth range of *Penicillium nalgiovense*. Preda and Kory (1988) reported that *Penicillium nalgiovense* is able to grow at pH values between 2 and 8.5, although optimal growth occurs at between pH 6.5 and 7.0.

5.5.3 Enzyme production

The proteolytic and lipolytic activity of yeasts such as *Debaryomyces hansenii*, *Hypopichia burtonii*, *Trichosporon cutaneum* and *Trichosporon pullulans* may be important in the curing and ripening of some hams and sausages (Comi and Cantoni, 1983) although Huerta *et al.* (1988) found that most of the yeasts isolated from dry-cured Spanish hams were lipolytic but not proteolytic. Molina *et al.* (1991) examined the lipolytic activity of *Cryptococcus albidus* when inoculated into a model meat system. Lipolytic activity was measured by the production of volatile and non-volatile acidity. Results suggested that *Cryptococcus albidus* has a much lower lipolytic activity than *Lactobacillus curvatus*, *Pediococcus pentosaceus* and *Staphylococcus xylosus* which are used in meat fermentations.

Comi *et al.* (1983) also found that the lipolytic activity of yeasts isolated from Parma ham was low. However, lipolytic activity by yeasts is thought to more significant in some sausage fermentations (Miteva *et al.*, 1986).

Although many genera of fungi have been screened for proteolytic activity, proteolytic enzymes have been characterised in relatively few *Penicillium* spp. Chrzanowska *et al.* (1993) examined production of proteases by a number of *Penicillium* species including *Penicillium chrysogenum* and *P. cyclopium*. In a starch/soya meal medium *P. chrysogenum* produced similar levels of proteases at pH 7.0 and pH 3.2 but under the same conditions proteolytic activity of *P. cyclopium* was much lower. However, in a glucose/casein medium at pH 3.2, the mould produced twice as much protease as *P. chrysogenum*. Moulds are likely to contribute to proteolytic activity in fermented meats although the activity of proteolytic enzymes does not appear to have been measured in meat substrates under realistic temperature, pH and a_w conditions.

Although the presence of moulds in some sausage fermentations contributes to flavour production, colour and chemical changes, the significance of moulds in the curing of raw ham is less clear. Huerta *et al.* (1987b) screened enzyme activities of 33 *Aspergillus* and 41 *Penicillium* strains isolated from Spanish dry-cured hams. Proteolytic activity was shown by 31% of *Aspergillus* and 20% of *Penicillium* strains. All *Aspergillus* strains, and 81% of *Penicillium* strains showed lipolytic activity. Toldrá and Etherington (1988) demonstrated that pork muscle proteinases (cathepsins B, D, H and L) were still active after eight months of curing for Serrano dry-cured hams. Plate counts of microorganisms suggested that numbers were insufficient to account for the enzyme activity in the interior of the ham. Since most fungal growth on hams occurs on the surface, the contribution of *Aspergillus*, *Eurotium*, *Penicillium* and yeasts to the flavour of the cured ham is likely to be minimal. The major role of muscle proteinases in the breakdown of muscle proteins during dry curing of hams was confirmed by Molina and Toldrá (1992) who found little evidence for significant proteolytic activity by *Pediococcus pentosaceus* and *Staphylococcus xylosus*.

5.5.4 Antimicrobials produced by fungi

The presence of antibiotic residues in foods is usually considered undesirable although many fermented foods contain microorganisms capable of producing a range of antimicrobial compounds which vary in activity, specificity and toxicity. Moulds are well known as producers of mycotoxins as well as a number of antibacterial agents (Laskin and Lechevalier, 1988). Antibiotics such as penicillin may be produced by *Penicillium chryso-*

genum, P. expansum, P. nalgiovense and other penicillia associated with fermented meats (Leistner and Ayres, 1968; Glenn *et al.*, 1989; Farber (unpublished) cited by Geisen, *et al.*, 1993). Metabolites from these moulds may prevent or reduce the growth and survival of undesirable Gram-positive bacteria such as *Staphylococcus aureus* and Glenn *et al.* (1989) concluded that *Penicillium nalgiovense* was able to form a substance which is inhibitory to *Listeria monocytogenes* in raw ripened sausages. In inoculation experiments, *Listeria monocytogenes* was added to an emulsion for raw ripened sausages and following casing the sausages were surface inoculated with *Penicillium nalgiovense*. Over a period of 22 days, the numbers of *Listeria monocytogenes* in the sausages were reduced by a factor of 100–400 near the casing and by a factor of 10 in the centre of the sausages.

Huerta *et al.* (1987b) examined the antimicrobial activity of *Aspergillus* and *Penicillium* strains isolated from Spanish dry-cured ham. *Eurotium amstelodami, E. chevalieri* and *E. halophilicus* showed antimicrobial activity towards *Escherichia coli*, *Proteus vulgaris*, *Salmonella typhimurium* and *Shigella sonnei* but not towards *Bacillus subtilis* or *Pseudomonas aeruginosa*. *Ecurotium halophilicus* also showed some activity against *Staphylococcus aureus*, *Saccharomyces cerevisiae* and *Candida albicans*. A notable finding was that *Penicillium* displayed more antimicrobial activity than *Aspergillus* or *Eurotium* species, particularly towards Gram-positive bacteria. *Penicillium* species dominate the early stages of mould ripening and antibacterial activity may be more important at this stage since the higher a_w will enable more bacterial growth to occur. No studies seem to have been conducted to examine the interaction between fungal metabolites and the bacterial flora during the early stages of mould growth.

5.6 Fungal starter cultures

5.6.1 Use of existing strains

Many producers of raw sausages and, in particular, raw hams still rely on the house flora to provide a natural inoculum for mould ripening. However, because a wide range of potentially toxinogenic moulds can grow on the surface of these meats there has been considerable interest in developing non-toxinogenic fungal starter cultures for ripening these products. Many *Aspergillus* and *Penicillium* species are capable of producing mycotoxins (Leistner and Ayres, 1968; Frisvad and Filtenborg, 1983, 1989; Leistner, 1984; Frisvad and Samson, 1991; see also Chapter 9) and there are relatively few strains which have received GRAS (Generally Regarded As Safe) status (Campbell-Platt and Cook, 1989). Even strains of *Penicillium camembertii*, *Penicillium roquefortii* and *Aspergillus oryzae* used in the production of fermented foods may be capable of producing toxic metabolites, a situation which is clearly undesirable in foods (Pitt *et al.* 1986; Samson and Van Reenen-Hoekstra, 1988; Leistner *et al.*, 1989; Engel and Teuber, 1990). An additional concern is the occurrence of potentially pathogenic fungi such as *Scopulariopsis* spp. since these have frequently been isolated from mould-ripened sausages and hams (Table 5.2).

Strains of *Penicillium nalgiovense* and *Penicillium chrysogenum* have been suggested as starter cultures for use in the ripening of meats. Many strains of *Penicillium chrysogenum* produce the mycotoxin roquefortine C (El-Banna *et al.*, 1987b) and antibiotics such as penicillin and negapillin (Schimi *et al.*, 1966; Laskin and Lechevalier, 1988). Although penicillin production could be beneficial in suppressing the growth of pathogenic bacteria it may also inhibit desirable Gram-positive bacteria such as *Pediococcus* and *Lactobacillus* species.

Pitt (1985), cited by Leistner *et al.* (1989) distinguished six morphological biotypes of *Penicillium nalgiovense* based on conidial colour, growth rate and conidial production. The colonies on agar varied from green/blue with good conidial production (biotypes 1, 4, 5) through to white with reduced conidial formation (biotypes 2, 3, 6). Biotypes 1–3 were found to form larger colonies than biotypes 4–6. The three strains which produced white conidia were considered, on the basis of their morphological appearance, to be suitable for use as starter cultures. However, not all strains of *Penicillium nalgiovense* are suitable as starter cultures because they may be capable of producing toxins or they may not impart the right characteristics to the fermented meat product (El-Banna *et al.*, 1987a; Fink-Gremmels *et al.*, 1987, 1988; Leistner *et al.*, 1989; Fink-Gremmels and Leistner, 1990).

Mintzlaff and Leistner (1972) were the first to select a non-toxinogenic and technologically suitable strain of *Penicillium nalgiovense* for meat ripening and this was subsequently used as a commercial starter culture known as 'Edelschimmel Kulmbach' (Mintzlaff and Christ 1973; Liepe, 1979, 1983, 1987). However, because of the variety of manufacturing methods, product types and consumer preferences, there is need for a variety of non-toxinogenic starter-culture strains. Moulds contribute many characteristics to fermented meats and lack of toxicity, although an important requirement, is not the only consideration in the selection of strains suitable for commercial purposes. Characteristics such as colour, growth rate, competitiveness, resistance to abrasion, ability to grow on different casings and enzyme production may influence the nature of the final product (Table 5.3 and Hwang *et al.*, 1993).

Fink-Gremmels *et al.* (1988) and Fink-Gremmels and Leistner (1990) showed that of 119 isolates of *Penicillium nalgiovense* in their collection, 50 isolates could be considered as potential starter cultures on the basis of

Table 5.3 Characteristics which influence the suitability of moulds for meat fermentations

Mould characteristic	Effect
Enzyme activity	
Proteases	Raise pH
Lipases	Contribute flavour or rancidity
Nitrate reductase	Enhance surface colour
Peroxidase	Reduce fat oxidation
Catalase	Breakdown peroxide
Cellulase	May destroy cellulose casings
Substrate utilisation	
Organic acids (lactate, acetate)	pH rises
Proteins	pH rises
Metabolite production	
Antibiotics (e.g. penicillin)	Inhibit pathogens and/or fermentative bacteria
Mycotoxins	Contaminate product
Volatiles	Desirable flavour and aroma or undesirable off-odours
Growth and sporulation	
Mycelium	Regulate moisture loss
	Reduce light, appearance, improve skin-peeling, resistance to abrasion
Spore production	Colour, flavour, uniformity
Strain stability	
Good competitiveness	Persist during ripening, suppress toxinogenic moulds
Antimicrobial sensitivity	Tolerate phenolic smoke components and spices
Mutations	Loss of desirable characteristics (colour, flavour, competitiveness, growth rate)
Fungal viruses	Reduced growth and sporulation

conidial colour and of these, 21 were suitable based on toxicological testing for 27 mycotoxins. As a result of assays using brine shrimp larvae and NMRI mouse screening tests, 36% of 112 *Penicillium nalgiovense* isolates were regarded as non-toxic. After evaluation of sausage ripening using 14 strains, only five were considered satisfactory in terms of lack of toxicity and morphological suitability by forming white colonies on sausages and laboratory media. These findings suggest that suitable starter-culture strains are far from common and it may be that genetic manipulation will offer the most practical route to develop strains with desirable characteristics.

Although mould starter cultures have been developed primarily for dry-sausage ripening, there is a need for suitable starter cultures for raw hams. Apart from safety considerations, uncontrolled ripening may result in defects such the growth of undesirable fungi which alter the appearance or lead to the production of taints such as phenol defects (Spotti *et al.*, 1988). *Penicillium nalgiovense* might not be suitable as a starter culture for raw hams since the ripening times may be up to two years and mycoflora changes from one initially dominated by *Penicillium* to one dominated by species of *Aspergillus* and *Eurotium*. A possible alternative would be to use a mixed starter culture perhaps consisting of *Penicillium nalgiovense* with *Eurotium repens* or *Eurotium rubrum*.

Less is known about the role of yeasts in meat fermentation although these fungi can tolerate the reduced a_w of these products and they may contribute to the organoleptic properties of both salami and dry-cured hams. Surface growths of *Debaromyces* and *Candida* spp. may consume oxygen, degrade peroxide, show lipolytic and, to a lesser extent, proteolytic activity, reduce moisture loss during curing and protect the meat from light. Both flavour and colour have been reported to be improved by the addition of selected *Debaromyces* strains to sausage mixtures (Rossmanith *et al.*, 1972; Coretti, 1977) and Vayssier (1979) suggested the use of *Debaryomyces hansenii* as a starter culture for ripening dry sausages, either alone, or in combination with a white mutant of *Penicillium nalgiovense*.

Fungal starter cultures are usually supplied as freeze-dried spores or liquid spore concentrates which, after adding to water, can be sprayed on the cased sausages. Alternatively, sausages can be dipped in vats of spores suspensions to provide an even distribution of inoculum. A high humidity is necessary to prevent drying out of the sausage casing since this would slow the growth of the starter culture. Hygienic conditions are important to avoid contamination by undesirable moulds particularly in those premises which formerly relied on inoculation by the house flora (Liepe, 1983; Le Bars and Le Bars, 1993). Hwang *et al.* (1993) reported that fungal starter culture spore suspensions have a shelf life of longer than nine months at $-28°C$.

5.6.2 Genetic manipulation of fungal starter cultures

The development of fungal starter cultures which are free of toxin production is an important consideration in fermented food production and the problem could be addressed by genetic manipulation. Mutation is one possible route but the results are unpredictable and the loss of an undesirable characteristic such as mycotoxin production could be accompanied by the development of undesirable traits such as reduced growth and enzyme production, unattractive colour or poor competitive ability.

Attention is now focusing on the application of molecular techniques to eliminate mycotoxins and to regulate fermentation characteristics such as proteolytic or lipolytic activity, and bio-preservative properties such as the production of antibiotics or hydrogen peroxide (Geisen, 1993). As with moulds of the genus *Aspergillus* the ability to express heterologous genes has now been demonstrated in *Penicillium nalgiovense* (Geisen *et al.*, 1988, 1990) Geisen *et al.* (1990) were able to successfully express the

Staphylococcus staphylolyticus lysostaphin gene in *Penicillium nalgiovense* and they found that some of the transformants secreted lysostaphin which, in *in vitro* tests, showed strong antimicrobial activity towards *Staphylococcus aureus*. Geisen (1993) has suggested that additional targets for genetic manipulation could include the removal of protease repression by low-molecular-weight nitrogen sources such as the nitrate added to fermented meats. Providing that the use of genetically manipulated strains is permitted in sausage manufacture and is accepted by consumers, then the technology should eventually lead to shorter fermentation and ripening times and a safer, more consistent product. It remains to be seen if the benefits of genetically manipulated strains can overcome legislative, manufacturing and consumer resistance to their introduction.

References

Alford, J.A., Pierce, D.A. and Suggs, F.G. (1964) Activity of microbial lipases on natural fats and synthetic triglycerides. *J. Lipid Res.*, **5**, 390–394.

Andersen, S.J. (1993) Potential mycotoxin producing moulds isolated from naturally mould-fermented sausages. Proceedings of meeting on Spoilage of Foods and Feeds. The Biodeterioration Society, Reading University, September 1992. *Int. Biodet. Biodeg.* **32**, 225.

Anon (1978) Fast aging with mould spray. *Meat Ind.*, August, 51.

Ayers, J.C., Leistner, L., Sutic, M., Koehler, P.E., Wu, M.T., Halls, N.A., Strzelecki, E. and Escher, F. (1974) Mould growth and mycotoxin production on aged hams and sausages *Proc. IVth Int. Congr. Food Sci. Technol.*, **3**, 218–227.

Besançon, X., Smet, C., Chabalier, C., Rivemale, M., Reverbel, J.P., Ratomahenina, R. and Galzy, P. (1992) Study of surface yeast flora of Roquefort cheese. *Int. J. Food Microbiol.*, **17**, 9–18.

Brooks, F.T. and Kidd, M. (1921) Black spot of chilled and frozen meat. *Special Report No. 6*. Food Investigation Board, London.

Brooks, F.T. and Hansford, C.G. (1923) Mould growth upon cold-store meat. *Trans. Br. Mycol. Soc.*, **8**, 113–141.

Campbell-Platt, G. (1987) *Fermented Foods of the World: A Dictionary and Guide.* Butterworths, London.

Campbell-Platt, G. and Cook, P.E. (1989) Fungi in the production of foods and food ingredients. *J. Appl. Bact. Symp. Ser.*, 119S–131S.

Cesari, E.P. and Guilliermond, A. (1920) The yeasts of sausages. *Annu. Rev. Inst. Pasteur* **34**, 229–233.

Chrzanowska, J., Kolaczkowska, M. and Polanowski, A. (1993) Production of exocellular proteolytic enzymes by various species of *Penicillium*. *Enzyme Microb. Technol.*, **15**, 140–143.

Comi, G. and Cantoni, C. (1983) Presenza di lieviti nei prosciutti crudi stagionati. *Ind. Alimentari*, February, 102–104.

Comi, G., Drago, G., Fagnani, V., Gaggero, L., Rossi, E. and Cantoni, C. (1983) Attività lipolytica di lieviti isolati da prosciutti crudi. *Conservazione degli Alimenti*, **9**, 12–16.

Cook, P.E. and Campbell-Platt, G. (1994) *Aspergillus* and fermented foods. In *Aspergillus*. Eds K.A. Powell, J Peberdy and A. Renwick. FEMS Symposium no 69, Plenum, New York, pp. 171–188..

Coretti, K. (1977) Starterkulturen in der Fleischwirtschaft. *Fleischwirtschaft.* **3**, 386–388.

Dalla Rosa, M., Manzano, M. and Cherubin, S. (1987) Gas chromatographic monitoring of induced fungal growth in cured ham. *Industrie Alimentari*, **26**, 545–549.

Deak, T. and Beuchat, L. (1987) Identification of foodborne yeast. *J. Food Protect.*, **50**, 243–264.

Dillon, V.M. and Board, R.G. (1991) Yeasts associated with red meats. *J. Appl. Bact.*, **71**, 93–108.

Dragoni, I., Ravenna, R. and Marino, C. (1980) Description and classification of *Aspergillus* species isolated from the surface of Parma and San Daniele raw-ripened hams. *L'Arch. Vet. Italiano*, **31**, Suppl. 5, 1–56.

Drake, S.D., Evans, J.B. and Niven, Jr. C.F. (1958) Microbial flora of packaged frankfurters and their radiation resistance. *Food Res.*, **23**, 291–296.

Dykes, G.A., Cloete, T.E. and Von Holy, A. (1991) Quantification of microbial populations associated with the manufacture of vacuum-packaged, smoked Vienna sausages. *Int. J. Food Microbiol.*, **13**, 239–248.

El-Banna, A.A., Pitt, J.I. and Leistner, L. (1987*a*) Production of mycotoxins by *Penicillium* species. *System. Appl. Microbiol.*, **10**, 42–46.

El-Banna, A.A., Frink-Gremmels, J. and Leistner, L. (1987*b*) Investigation of *Penicillium chrysogenum* isolates for their suitability as starter cultures. *Mycotox. Res.*, **3**, 77–83.

Ellis, J.J., Wang, H.L. and Hesseltine, C.W. (1974) *Rhizopus* and *Chlamydomucor* strains surveyed for milk-clotting, amylolytic and antibiotic activities. *Mycologia*, **66**, 593–599.

Engel, G. and Teuber, M. (1990) Toxic metabolites from fungal cheese starter cultures (*Penicillium camemberti* and *Penicillium roqueforti*). In *Mycotoxins in Dairy Products*. eds H.P. Van Egmond, pp. 163–192. Elsevier Science Publishers, London.

Fielder, H. (1973) Schimmelpilzdiagnostik und Mykotoxinnachweis an fleischprodukten. *Arch. Lebensmittelhygiene*, **8**, 180–184.

Fink-Gremmels, J. and Leistner, L. (1990) Toxicological evaluation of moulds. *Food Biotechnol.*, **4**, 579–584.

Fink-Gremmels, J., El-Banna, A.A. and Leistner, L. (1987) Toxicological evaluation of mould starter cultures. *Mitteilungsbl. Bundesanstalt Fleischforsch., Kulmbach*, No. 105, 317–324.

Fink-Gremmels, J., El-Banna, A.A. and Leistner, L. (1988) Developing mould starter cultures for meat products. *Fleischwirtschaft.* **68**, 1292–1294.

Frisvad, J.C. (1988) Fungal species and their specific production of mycotoxins. In *Introduction to Food-borne Fungi*. 3rd edn. eds R.A. Samson and E.S. van Reenen-Hoekstra, CBS Baain, pp. 239–249.

Frisvad, J.C. and Filtenborg, O. (1983) Classification of terverticillate penicillia based on profiles of mycotoxins and other secondary metabolites. *Appl. Environ. Microbiol.*, **46**, 1301–1310.

Frisvad, J.C. and Filtenborg, O. (1989) Terverticillate penicillia: Chemotaxonomy and mycotoxin production. *Mycologia*, **81**, 837–861.

Frisvad, J.C. and Samson R.A. (1991) Filamentous fungi in foods and feeds: Ecology, spoilage, and mycotoxin production. In *Handbook of Applied Mycology*, Vol. 3, *Foods and Feeds*. Eds D.K. Arora, K.G. Mukerji and E.H. Marth, pp. 31–68. Marcel Dekker, New York.

Geisen, R. (1993) Fungal starter cultures for fermented foods: molecular aspects. *Trends Food Sci. Technol.*, **4**, 251–256.

Geisen, R., Glenn, E. and Leistner, L. (1988) Development of a transformation system for *Penicillium nalgiovense*. *Proc. Japan. Assoc. Mycotox.*, Suppl. 1, 47–48.

Geisen, R., Ständner, L. and Leistner, L. (1990) New mould starter cultures by genetic modification. *Food Biotechnol.*, **4**, 497–503.

Gill, C.O. and Lowry, P.D. (1982) Growth at subzero temperatures of blackspot fungi from meat. *J. Appl. Bacteriol.*, **52**, 245–250.

Gill, C.O., Lowry, P.D. and Di Menna, M.E. (1982) A note on the identities of organisms causing black spot spoilage of meat. *J. Appl. Bacteriol.*, **51**, 183–187.

Glenn, E., Geisen, R. and Leistner, L. (1989) Control of *Listeria monocytogenes* in mould ripened raw sausages by strains of *Penicillium nalgiovense*. *Mitteilungsbl. Bundesanstalt Fleischforsch. Kulmbach* No. 105, 317–324.

Grazia, L., Romano, P., Bagni, A., Roggiani, D. and Guglielmi, G. (1986) The role of moulds in the ripening process of salami. *Food Microbiol.*, **3**, 19–25.

Haines, R.B. (1931) The growth of microorganisms on chilled and frozen meat. *J. Soc. Chem. Ind. (Lond.)*, **50**, 223T–227T.

Hadlok, D., Samson, R.A. and Schnorr, B. (1975) Moulds and meat: The genus *Penicillium. Fleischwirtschaft*, **55**, 979–983.

Hadlok, D., Samson, R.A. and Schipper, M.A.A. (1976) Schimmelpilze und Fleisch Kontaminationsflora. *Fleischwirtschaft*, **56**, 372–376.

Holley, R.A. (1986) Effect of sorbate and pimaricin on surface mould and ripening of Italian dry salami. *Lebensmittel-Wissenschaft Technol.*, **19**, 59–65.

Huerta, T., Sanchis, V., Hernandez-Haba, J. and Hernandez, E. (1987a) Mycoflora of dry salted Spanish ham. *Microbiologie–Aliments–Nutrition*, **5**, 247–252.

Huerta, T., Sanchis, V., Hernandez-Haba, J. and Hernandez, E. (1987b) Enzymic activities and antimicrobial effects of *Aspergillus* and *Penicillium* strains isolated from Spanish dry cured hams. *Microbiologie–Aliments–Nutrition*, **5**, 289–294.

Huerta, T., Quenol, A. and Hernandez-Haba, J. (1988) Yeasts of dry cured hams, qualitative and quantitative aspects. *Microbiologie–Aliments–Nutrition*, **6**, 289–294.

Hwang H.-J., Vogel, R.F. and Hammes, W.P. (1993) Development of mould cultures for sausage fermentation. Technological properties of the strains and sensorial assessment of the products. *Fleischwirtschaft*, **73**, 327–328, 331–332.

Ingram, M. and Dainty, R.H. (1971) Changes caused by microbes in spoilage of meat. *J. Appl. Bact.*, **34**, 21–39.

Jay, J.M. (1978). Meat, poultry and seafoods. In *Food and Beverage Mycology* Ed. L.R. Beuchat, pp. 155–173. AVI Publishing, New York.

Jay, J.M. and Margitic, S. (1981) Incidence of yeasts in fresh ground beef and their ratios to bacteria. *J. Food Sci.*, **46**, 648–649.

Katsaras, K. and Budras, K.-D. (1992) Microstructure of fermented sausage. *Meat Sci.*, **31**, 121–134.

Kemp, J.D., Langlois, B.E. and Fox, J.D. (1983) Effect of potassium sorbate and vacuum packaging on the quality and microflora of dry-cured, intact and boneless hams. *J. Food Sci.*, **48**, 1709.

Kotula, A.W., Campano, S.G. and Kinsman, D.M. (1982) Proteolytic and lipolytic activity of moulds isolated from aged beef. *J. Food Protect.*, **45**, 1242–1244, 1247.

Laskin, A.I. and Lechevalier, H.A. (1988) *Handbook of Microbiology*, Vol. IX, Part A. *Antibiotics*. Eds A.I. Laskin H.A. and Lechevalier, 2nd ed. CRC Press, Boca Raton.

Le Bars, J. and Le Bars, P. (1993) Contamination fongique de salaisons seches de viande: Origine, conditions d'apparition, prevention. *Cryptogamie Mycol.*, **13**, 193–202.

Leistner, L. (1984) Toxigenic penicillia occurring in feeds and foods: a review. Food Technol. Austral., **36**, 404—406, 413.

Leistner, L. (1986) Mould-ripened foods. *Fleischwirtschaft*, **66**, 1385–1388.

Leistner, L. and Ayres, J.C. (1968) Moulds and meat. *Fleischwirtschaft*, **48**, 62–65.

Leistner, L. and Bem, Z. (1970) Vorkommen und Bedeutung von Hefen bel Pokelfleischwaren. *Fleischwirtschaft*, **50**, 350–351.

Leistner, L. and Wirth, F. (1972) Importance and determination of water activity (a_w value) of meat and meat products. *Fleischwirtschaft*, **52**, 1335–1337.

Leistner, L., Geisen, R. and Fink-Gremmels, J. (1989) Mould-fermented foods of Europe: Hazards and developments. In *Mycotoxins and Phycotoxins '88*. Eds S. Natori, K. Hashimoto and Y. Ueno, pp. 145–154, Elsevier, Amsterdam.

Liepe, H.-U. (1979) Mould culture — a contribution to solution of the mycotoxin problem. *RTVA* **18**, 48–49.

Liepe, H.-U. (1983) Starter cultures in meat production. In *Biotechnology*, Vol. 5. Ed. G. Reed, pp. 399–424. Verlag Chemie, Weinheim.

Liepe, H.-U. (1987) Manufacture of air-dried raw sausage with mould culture. *Fleischerei*, **38**, 546–547.

Leon, F., Martins, C., Mata, C., Penedo, J.C., Barranco, A., Camargo, S., Martinez, I., Velloso, C., Jorquera, D., Torres, J.M. and Moreno, R. (1988) pH value and water activity in Iberian Serrano ham. Repercussions due to the EEC. *Alimentaria*, **189**, 9–11.

Lowry, P.D. and Gill, C.O. (1984) Temperature and water activity minima for growth of spoilage moulds from meat. *J. Appl. Bact.*, **56**, 193–199.

McCarthy, J.A. and Damoglou, A.P. (1993) The effect of low-dose gamma irradiation on yeasts of British fresh sausage. *Food Microbiol.*, **10**, 439–446.

Michener, H.D. and Elliot, R.P. (1964) Minimum growth temperatures for food poisoning faecal indicators and psychrophilic microorganisms. *Adv. Food Res.*, **13**, 349–396.

Mintzlaff, H.-J. and Leistner, L. (1972) Untersuchungen zur Selektion eines technologisch geeigneten und toxikologisch unbedenklichen Schimmelpilz-stammes für die Rohwurstherstellung. *Zentral Bl. Vet. Med.* B, **19**, 291–300.

Mintzlaff, H.-J. and Christ, W. (1973) *Penicillium nalgiovensis* as starterculture for Suedtiroler Bauernspeck. *Fleischwirtschaft*, **53**, 864–867.

Miteva, E., Kirova, E., Gadjeva, D. and Radeva, M. (1986) Sensory aroma and taste profiles of raw dried sausages manufactured with a lipolytically active yeast culture. *Nahrung*, **30**, 829–832.

Molina, I. and Toldrá, F. (1992) Detection of proteolytic activity in microorganisms isolated from dry-cured ham. *J. Food Sci.*, **57**, 1308–1310.

Molina, I., Silla, H. and Flores, J. (1990) Study of the microbial flora in dry-cured ham. 4. Yeasts. *Fleischwirtschaft*, **70**, 74–76.

Molina, I., Nieto, P., Flores, J., Silla, H. and Bermell, S. (1991) Study of the microbial flora in dry-cured ham. 5. Lipolytic activity. *Fleischwirtschaft*, **71**, 906–908.

Monte, E., Villanueva, J.R. and Dominguez, A. (1986) Fungal profiles of Spanish country-cured hams. *Int. J. Food Sci. Technol.*, **3**, 355–359.

Moreau, C. (1979) *Moulds, Toxins and Foods*. John Wiley, London.

Mrak, E.M. and Bonar, L. (1938) A note on yeast obtained from slimy sausage. *Food Res.* 3, 615–618.

Pitt, J.I., Cruickshank, R.H., Leistner, L. (1986) *Penicillium commune*, *P. camembertii*, the origin of white cheese moulds, and the production of cyclopiazonic acid. *Food Microbiol.*, **3**, 363–371.

Preda, N. and Kory, M. (1988) Verhalten von *Penicillium nalgiovensis* unter verschiedenen Bedingungen. Daten für die Herstellung schimmelpilzgeneifter Rohwürste. *Fleischwirtschaft*, **68**, 1015–1017.

Rojas, F.J., Jodral, M., Gosalvez, F. and Pozo, R. (1991) Mycoflora and toxinogenic *Aspergillus flavus* in Spanish dry-cured ham. *Int. J. Food Microbiol.*, **13**, 249–256.

Roncalés, P., Aguilera, M., Beltran, J.A., Jaime, I. and Peiro, J.M. (1991) The effect of natural or artificial casing on the ripening and sensory quality of a mould-covered dry sausage. *Int. J. Food Sci. Technol.*, **26**, 83–89.

Rossmanith, E., Mintzlaff, H-J., Streng, B., Christ, W. and Leistner, L. (1972) Hefen als Starterkulturen für Rohwürste. *Jahresben. Bundesanstalt Fleishforsch.*, 147–148.

Samson, R.A. and van Reenen-Hoekstra, E.S. (1988) *Introduction to Food-borne Fungi*, 3rd edn. Centraalbureau voor Schimmelcultures, Baarn.

Schimi, I.R., Iman, G.M. and Saad, A. (1966) Negapillin, a new antibiotic isolated from metabolites of *Penicillium chrysogenum*. *J. Antibiot. Tokyo*, Series A, **19**, 19–22.

Schmidt-Lorenz, W. and Gutschmidt, J. (1969) Mikrobielle und sensoriche Veränderungen gefrorener Brathahnchen und Poularden bei Lagerung in Temperaturbereich. *Fleischwirtschaft*, **49**, 1033–1041.

Scott, W.J. (1936) The growth of microorganisms on ox muscle. The influence of water content of substrate on rate of growth at −1°C. *J. Counc. Sci. Ind. Res.*, **9**, 177–190.

Semeniuk, G. and Bell, W.C. (1937) Some moulds associated with meat in cold storage lockers in Iowa. *Proc. Iowa Acad. Sci.*, **44**, 37–43.

Skrinjar, M. and Horvat-Skenderovic, T. (1989) Contamination of dry sausage with moulds, aflatoxins, ochratoxins and zearalenone. *Technologija Mesa*, **30**, 53–59.

Smith, M. Th. and Hadlock, R. (1976) Hefen und Fleisch: Voorkomen, Systematik und Differezierung. *Fleischwirtschaft*, **56**, 379–384.

Spotti, E., Mutti, P. and Campanini, M. (1988) Microbiological study of the 'phenol defect' of ham during ripening. *Industria Conserve*, **63**, 343–346.

Toldrá, F. and Etherington, D.J. (1988) Examination of cathepsins B, D, H and L activities in dry-cured hams. *Meat Sci.*, **23**, 1–7.

Vayssier, Y. (1979) New mould culture for surface ripening of dry sausage. *RTVA*, **18**, 50.

Viljoen, B.C., Dykes, G.A., Callis, M. and von Holy, A. (1993) Yeasts associated with Vienna sausage packaging. *Int. J. Food Microbiol.*, **18**, 53–62.
Williams, A.P. (1990) *Penicillium* and *Aspergillus* in the food microbiology laboratory. In *Modern Concepts in* Penicillium and Aspergillus *Classification*. Eds R.A. Samson and J.I. Pitt, pp. 121–137, Plenum Press, New York.
Williams, B.E. (1957) Method for aging meat. US Patent No. 2, 816, 836.
Williams, B.E. (1962) Methods for aging and flavoring meat. US Patent No. 3, 056, 679.

6 Starter cultures for meat fermentations

B. JESSEN

6.1 Introduction

The origin of fermented meat products still remains unknown, although the history of fermented sausages can probably be dated back more than 2000 years to the Mediterranean (Pederson, 1979). The name salami may be derived from the destroyed city Salamis on the east coast of Cyprus (Pederson, 1979). However, Liepe (1983), who refers to the Greek island Salamis, strongly denies this and advocates that the name originates from the Italian word for salt, *sale*. Finally, Pederson (1979) also suggests that the word sausage originates from the Latin word *salsus*.

In spite of the differences of opinion in dating the history of fermented meat, it is a fact that today's sausages are very much based on the ancient type of sausage. Like then, sausages are nowadays based on comminuted meat, mixed with salt and spices, and stuffed into casings, ripened and dried. The sausages may be surface-treated, i.e. smoked or moulded. The industrial development in the second half of the nineteenth century led to the use of starter cultures for interior as well as exterior use.

Such a simple description of dried-sausage manufacturing can be used to characterize fermented sausages throughout the world, although a complex variety of different products exists. National or local differences arise in the choice of recipes, seasonings, type of starter culture, degree of comminution, calibre size, fermentation conditions and surface treatment. Southern European types are often heavily spiced whereas the Northern types are less spicy, heavily smoked and rather salty. The varieties have often been named after the cities or areas of origin (Pederson, 1979).

In principle, excellent salami can be produced without use of starter cultures. Traditionally, the chopped meat was presalted in order to promote development of lactic acid bacteria or the 'back-slopping' method was used. However, large-scale industrial production requires uniform and accelerated processing which has led to the present general use of starter cultures.

The first commercially available starter culture was marketed in the USA in 1957. It was a single strain culture of *Pediococcus cerevisiae* (Niven *et al.*, 1959), later renamed *Pediococcus acidilactici* (Everson *et al.*, 1970), to be used for US types of sausages like summer sausages, cervelat and thuringer. The culture was marketed under the brand name 'Accel' by

Merck. In contrast, the first commercialised European starter culture was a single-strain *Micrococcus* culture (strain name M53) which was introduced in 1961 by the German company Rudolf Müller under the name of 'Baktoferment 61' (Niinivaara *et al.*, 1964).

The reason for coming up with two species from genera as far from each other as *Pediococcus* and *Micrococcus* is the difference in sausage technology within the two continents. In order to improve the speed of fermentation the US industry regards a fast acidification as the most important quality, and fermentations are carried out at temperatures around 38–40°C. As pediococci do not reduce nitrate, nitrite was used as the curing agent together with the starter culture resulting in a decrease in total processing time from six to two days (Everson *et al.*, 1970). In Europe, sausages were manufactured at ambient temperatures, i.e. around 20°C. The lower fermentation temperature has a positive impact on the colour and flavour development. Therefore, those micrococci which actively reduce nitrate and ferment carbohydrates slowly were the chosen bacteria in Europe (Niinivaara, 1955; Niinivaara *et al.*, 1964).

Originally, meat starter cultures were marketed as lyophilised cultures. In 1968, *Pediococcus acidilactici* was offered as a frozen concentrate in order to shorten the processing time. Hereby, the US producers could eliminate the traditional rehydration step and the processing time was reduced by another 12–16 hours (Everson *et al.*, 1970). Today, meat starter cultures are supplied both as frozen and as lyophilised products.

Since a modest start in the late 1950s, the market for meat starter cultures increased considerably around 1980, and today they have become a natural component of the sausage ingredients. The market has developed into mixed cultures consisting of strains which cover a wide variety of activities and growth tolerances.

6.2 Meat fermentation

The term 'fermented meat' is normally used to describe dry-cured sausages and ham products. Presently, the application of starter cultures is almost exclusively confined to fermented dried sausages.

In spite of this, experiments have revealed that starter cultures can also be advantageous for cooked sausages. Roca and Incze (1989) describe a positive effect of *Lactobacillus* on the taste of sausages which had been fermented for three days prior to heat treatment. Petäjä (1977) published results that showed improved effects on colour and flavour of frankfurters fermented one day with *Staphylococcus carnosus* before cooking. However, it has never become common to use starter cultures in cooked sausages.

In smeared sausages such as the German *streichfähige Rohwurst* the overall sensoric quality is improved by the use of starter cultures (Klettner and Lücke, 1992). Furthermore, an effective pH decline is ensured, thereby minimising the growth of undesirable microorganisms.

6.2.1   Sausage fermentation

Fermented dried sausages can be divided into three groups depending on the fermentation conditions.

Fermentation at high temperature.   Pepperoni and the US summer sausage are representatives of these types of sausages. Pepperoni is a dry sausage with a moisture : protein ratio of 1.6:1.0 and summer sausage a semi-dry sausage with a ratio of 2.0–3.7:1.0 (Bacus, 1984). The fermentation temperature is 40°C and the fermentation time is 15–20 hours followed by a hot rinse and heat treatment to a core temperature of 60°C. The fermentation can also be followed by a drying period.

The starter culture used is typically *Pediococcus acidilactici* which is characterised by a high-temperature optimum, i.e. 40°C.

Fermentation at traditional European temperatures.   In continental Europe, the temperature range for fermentation is 20–24°C with drying temperatures between 15 and 18°C. Optimal results can only be obtained with a climate chamber in which temperature and humidity can be controlled completely.

For this type of product, mixed cultures consisting of lactic acid bacteria (*Lactobacillus* or *Pediococcus*) and Micrococcaceae (*Staphylococcus* or *Micrococcus*) are usually used. Most strains belong to the mesophilic group although in recent years more psychrotrophic strains have appeared amongst the commercial starters.

Fermentation at low temperature.   In many countries, controlled climate chambers are still not common. In eastern and southern Europe, the fermentation is based on the natural climatic conditions, and therefore the risk of faulty fermentations is rather high.

Potential cultures for these markets consist of psychrotrophic lactic acid bacteria which are capable of growing at low temperatures and, therefore, perform well at temperatures between 10 and 15°C. Today, such cultures are commercially available.

6.2.2   Ham fermentation

Traditional dry-cured ham is produced by rubbing salt onto the surface of the pieces of meat which are stored at chilled temperatures, i.e. lower than 5°C. The salt is washed off and the meat may be resalted for another period. Then, the temperature is raised to 15–20°C for drying. The last step is the curing period, during which the ham is frequently covered with lard in order to prevent excessive drying out. The temperature may vary from 15 to 25°C depending on the season. A natural mould covering is characteristic for this part of the process.

The time for the various steps defines the quality of the ham. High-quality ham requires a processing time of up to two years. The famous ham types are often named for the place of origin, e.g. Parma ham and Ardenner ham.

Accelerated production processes for cured ham have been developed. In such cases the use of starter cultures is an advantage. The procedure is changed as the dry salting is replaced by brine salting or brine injection pumping. The starter culture, primarily salt-tolerant staphylococci or combinations of staphylococci and lactic acid bacteria, is added via the saline brine in order to stabilise and improve colour and flavour.

6.2.3 Cured-meat flavour components

To a certain extent the flavour and taste of cured-meat products are defined by ingredients such as salt, nitrite/nitrate and spices but also by choice of process, e.g. smoking that influences the overall sensory quality. In addition, it is described that several flavour components originate from the microbial decomposition of carbohydrates, lipids and proteins (Lücke, 1986a).

An early publication by Keller (1954) demonstrates that various Gram-negative rods exert a positive influence on the aroma development of dry sausages when the temperature is kept as low as 12°C. In spite of this, Gram-negative bacteria have for several reasons never played an important role in the fermentation of sausages. At higher temperatures, they may spoil the meat because of a strong proteolytic activity, and in addition, some are known as potential pathogens. Moreover, from a culture manufacturing point of view, their survival during freeze-drying is generally poor (Lücke and Hechelmann, 1987).

In a detailed review on cured meat flavour, Gray and Pearson (1984) bring up the question as to whether a 'cured flavour' exists or not. They state that although nitrite is associated with the particular cured flavour, the underlying chemistry behind this unique phenomenon is not entirely understood. The review concludes that the reactive nitrite may interact to result in unidentified cured components and also, that cured meat contains a relatively lower amount of carbonyls.

In a recent study on cured ham, Berdagué et al. (1991) found that several aromatic molecules could be related to the characteristic cured flavour. The compounds were detected by flavour tests and do, however, still need to be chemically identified.

6.3 Starter cultures

The microorganisms used for meat fermentation consist of a group of lactic acid bacteria (*Lactobacillus, Pediococcus*) plus the Micrococcaceae (*Staphylococcus, Micrococcus*), *Streptomyces, Debaryomyces* and *Penicillium*.

6.3.1 Lactic acid bacteria

Lactobacillus and *Pediococcus* are the two genera used for acidification of fermented sausages. Even though a very high number of preparations are commercially available, the number of species is limited. According to Hammes *et al.* (1985) only five species are in use: *Lactobacillus plantarum*, *Lactobacillus sake, Lactobacillus curvatus, Pediococcus pentosaceus* and *P. acidilactici*. In addition to these, *Lactobacillus pentosus* is found in some starter culture preparations (Rudolf Müller; Chr. Hansen).

The psychrotrophic strains *L. sake* and *L. curvatus* are technically more suitable to compete at traditional ripening temperatures but because of their tendency to produce hydrogen peroxide, the mesophilic *Pediococcus* species and *L. plantarum* are the preferred organisms (Lücke and Hechelmann, 1987). Pediococci are rarely isolated as part of the indigenous flora of sausages, but due to their ability to survive during lyophilisation, they are frequently used and have become almost the only species used on the US market (Hammes *et al.*, 1990). Nevertheless, even if pediococci are added to the meat at adequate levels, they tend to die off during fermentation. A phenomenon that may cause sensory faults is the growth of undesirable lactic acid bacteria among the spontaneous flora at the expense of the starter culture (Hammes *et al.*, 1990).

Acidification. The lactic acid bacteria used as meat starter cultures are preferably homofermentative. They decompose glucose and hexose phosphates with gluco-configuration through the Embden–Meyerhof–Parnas (EMP) pathway with lactate as the sole end-product (Kandler, 1983; Lücke and Hechelmann, 1987). Glucose is degraded into two pyruvate moieties which are further converted into lactate by lactate dehydrogenase. Lactic acid bacteria produce either DL−, D(−) or L(+)−lactic acid. Some organisms produce D(−)−lactic acid exclusively, whereas organisms producing L (+)-lactic acid always form smaller amounts of D(−)-lactic acid (Gottschalk, 1985).

Due to a decreased water-binding capacity of the meat proteins, the acidification accelerates drying out, and thus shortens the processing time. The flavour of sausages, fermented with a single strain of lactic acid bacteria, is characterised by a sharp tangy taste caused by lactic acid (Lücke, 1986a). In order to obtain a sufficient acidification, sausage recipes should always contain glucose, which is easily metabolised. In addition, the recipes may also include more complex carbohydrates such as dextrin, corn syrup and starches. However, the decomposition of these carbohydrates is slow and has no practical significance for the acidification if simple carbohydrates are present (Bacus and Brown, 1985).

Lactobacillus. Even though homofermentative lactobacilli are preferred for use in sausage manufacturing, all commercially available cultures belong to the group of facultatively heterofermentative lactobacilli (Kandler and Weiss, 1986). In addition to the enzyme aldolase necessary in the EMP pathway (glycolysis), they also possess phosphoketolase which decomposes pentoses leading to formation of lactate, acetate and carbon dioxide. Usually the carbohydrates are metabolised via glycolysis. However, under certain conditions, the heterofermentative pathway is activated, resulting in undesirable flavour components. Acetate belongs to this group.

Kandler (1983) has given a detailed description of the carbohydrate metabolism of *Lactobacillus*. Most lactic acid bacteria have specific permeases for the uptake of saccharides and oligosaccharides. Inside the cell, glycosidases split the oligosaccharides, and finally the monosaccharides are phosphorylated and metabolized through the EMP pathway.

Lactose is one of the disaccharides frequently used in sausage recipes. Lactobacilli take up lactose which is split into galactose and glucose by means of β-galactosidase. Galactose is converted into glucose 6-phosphate through the Leloir pathway and together with glucose fermented via glycolysis.

Lactobacillus plantarum is reported to oxidise lactic acid into acetate and carbon dioxide under aerobic conditions (Kandler, 1983). Borch and Molin (1989) have shown that on average, homofermentative lactobacilli under aerobic conditions produce 1.1 mole lactic acid, 0.4 mole acetic acid and minor amounts of acetoin from 1 mole glucose. The results of both studies indicate that off-flavour may appear in the final product. Other meat starters such as the psychrotrophic *L. curvatus* and *L. sake* use oxygen to form hydrogen peroxide (H_2O_2) and pyruvate of which the former increases the risk of discoloration (Kandler, 1983).

It is presumed that superoxide (O_2^-) is an intermediate in the H_2O_2 formation. One way of avoiding the scavenging effect of O_2^- is the action of superoxide dismutase. However, it has been demonstrated that *L. plantarum*, for example, does not contain this enzyme. Instead, the lactic acid bacteria use Mn^{2+} as a O_2^- scavenger and this may explain the very high manganese requirements since the manganese-dependent enzymes are saturated at much lower concentrations (Kandler, 1983). Raccach (1985) lists the enzyme systems stimulated by manganese. Among these are pyruvate kinase and other enzymes in the EMP pathway plus enzymes in the heterofermentative pathway. Manganese in meat systems has also been demonstrated to lead to an accelerated pH drop (Raccach and Marshall, 1985).

Even in anaerobic environments, it has been shown that *Lactobacillus* may switch from a homofermentative to a heterofermentative metabolism upon glucose depletion (Borch *et al.*, 1991). As a consequence, end-products change from lactate to formate and acetate and, in addition, small amounts of ethanol are observed.

Further, glucose depletion leads to a change in amino acid profiles. The decrease in sulphur-containing amino acids results in a concurrent increase in total sulphide (Borch *et al.*, 1991). All are reactions that may influence the flavour profile of the sausages.

High salt concentrations, i.e. 6–8% NaCl in the aqueous phase, as can be found in dry sausages, may alter the fermentation pattern (Bobillo and Marshall, 1991). Under aerobic conditions, *L. plantarum* produces less acetic acid and more lactic acid, whereas high salt concentrations under anaerobic conditions do not change the ratio between lactic acid and acetic acid but merely reduce the total acid production.

Pediococcus.   Like *Lactobacillus*, the main reason for the use of pediococci in meat products is the homofermentative production of lactate. Considering the importance of *Pediococcus* in food applications, the few reported data on basic physiology of pediococci is striking.

Results reported by Blickstad and Molin (1981) show that the lactic acid production of *Pediococcus pentosaceus* is directly linked to growth. Further, the strain is capable of growing within a varied environment of relatively wide limits. It has a high tolerance towards nitrite, aerobic and anaerobic atmospheres and works well over a broad pH range. All are valuable features with respect to meat fermentation. However, temperatures of 20°C or below result in a marked drop in growth rate.

Tetlow and Hoover (1988) demonstrated that *Pediococcus pentosaceus* produces not only lactate but also ethanol and acetate from hexoses and pentoses under non-limiting conditions. In particular, the amount of ethanol was unexpectedly high in hexose fermentation. This result may explain why pediococci are often observed to give a less tangy taste than lactobacilli when used in fermented sausages.

Nitrate and nitrite reductase.   The enzyme system necessary for the cured colour is normally attributed to Micrococcaceae. However, Hammes *et al.* (1990) summarise activities of lactic acid bacteria among which are mentioned nitrate and nitrite reductases. The latter can be either heme dependent or heme independent. Heme is not a limited component in meat and, therefore, lactic acid bacteria may also support the colour development of fermented meat.

Strains within *L. plantarum* and *L. pentosus* possess nitrate and nitrite reductase and some strains of *Pediococcus pentosaceus* possess the heme-dependent nitrite reductase. Nitrite reductase could not be demonstrated in *L. curvatus* and it was rather poor in *L. sake* (Wolf and Hammes, 1988).

The heme-dependent nitrite reductase results in ammonia as the end-product of nitrite whereas the heme-independent nitrite reductase reduces nitrite to nitrous oxide and nitric oxide, of which the latter contributes to colour formation. However, practical experiments have revealed that colour formation, based solely on *Lactobacillus* possessing nitrite reductase activity, was slow and insufficient, resulting in colour defects (Hammes *et al.*, 1990).

Catalase.   The catalase activity in meat starter cultures is also usually attributed to Micrococcaceae. However, several reports document that not only is catalase produced by lactic acid bacteria, they also form hydrogen peroxide under aerobic conditions (Whittenbury, 1964; Lücke *et al.*, 1986; Condon, 1987; Wolf and Hammes, 1988).

Lactic acid bacteria, generally, do not synthesise heme and, therefore, are devoid of heme-containing enzymes such as cytochromes, catalase and peroxidases. However, under aerobic conditions, oxidation can be mediated by flavoprotein oxidase enzymes with hydrogen peroxide or water as end-products (Condon, 1987). The detrimental effect of hydrogen peroxide on lactic acid bacteria is overcome by the presence of pseudo-catalase, a peroxidase, which is by nature a flavoprotein.

Under certain conditions, the heme-containing catalase may also be produced by lactobacilli and pediococci (Whittenbury, 1964), *Lactobacillus plantarum, L. sake* and *L. bavaricus* but not *L. curvatus* have been demonstrated to decompose hydrogen peroxide by means of the heme-containing catalase if the substrate contains hematin (Whittenbury, 1964; Wolf and Hammes, 1988), *Lactobacillus plantarum, L. sake* and *L. curvatus* produce hydrogen peroxide under aerobic conditions (Whittenbury, 1964; Lücke *et al.*, 1986; Borch and Molin, 1989) but as the last species does not have the capacity of producing catalase, it may cause greyish surface discoloration in sausage fermentation (Hammes *et al.*, 1990).

According to Whittenbury (1964), strains of *Pediococcus* and *P. acidilactici* produce hydrogen peroxide on certain substrates. Most *P. pentosaceus* strains decompose hydrogen peroxide by means of pseudocatalase whereas *P. acidilactici* possesses the heme-dependent catalase.

Aroma formation.   In meat fermentation, the contribution of lactic acid bacteria to aroma formation is usually limited to the tanginess of dry sausages (Lücke, 1986a). The tangy taste is a consequence of the lactic acid produced, possibly in connection with small amounts of acetic acid. As already mentioned, the environmental conditions may lead to a changed fermentation pattern resulting in end-products that influence the aroma formation. Too much acetic acid will result in an undesirable biting acidity whereas acetoin influences positively, imparting a nutty flavour and aroma (Pederson, 1979).

The lipolytic and proteolytic capacities of lactic acid bacteria in meat fermentation are considered to be of limited value. Several reports document such capacities although it may be difficult to connect it to cured flavour.

Lipolytic activity of lactic acid bacteria on synthetic media and natural fats have been reported by Coretti (1965), Reuter (1975), El Soda *et al.* (1986a), Papon and Talon (1988) and Nielsen and Kemner (1989). The lipolytic enzymes are intracellular and their production is correlated with the exponential growth phase and is stimulated by low glucose concentrations (Talon and Papon, 1988). The maximum lipase production is obtained at the optimum growth temperature (Papon and Talon, 1988; Nielsen and Kemner, 1989) and at neutral pH (El Soda *et al.*, 1986b; Papon and Talon, 1988). According to Nielsen and Kemner (1989), tributyrin cannot be used as test medium for lipolytic activity because the starter strains examined express activity on animal fat but not on tributyrin.

Lactic acid bacteria isolated from meat and meat products possess proteolytic activity when examined in culture media and sausages. Reuter (1975) observed a slight proteinase activity and also a considerable increase in free amino acids in culture medium and meat suspension. In sausages, Vignolo *et al.* (1988) show an optimal proteolytic activity in sausages at 40°C. The enzymatic activity is affected by the salt concentration, i.e. an increase in NaCl concentration from 3 to 5% reduces the proteolytic activity by 80%.

Næs *et al.* (1991) studied the effect of a commercial esterase and a proteinase, the latter being purified from *L. sake*, in a dry sausage production. Both influenced the sensory characteristics of the sausages significantly. Off-odour and bitter taste became more pronounced under the conditions examined.

Papon and Talon (1988) together with Nielsen and Kemner (1989) and Næs *et al.* (1991) conclude that lactic acid bacteria contain lipolytic enzymes that are active at temperatures and pH values relevant for sausage fermentation and that these enzymes may have an influence on the aroma. It has been shown that *L. curvatus* and *P. pentosaceus*, both isolated from dry-cured ham, also have lipolytic activity (Nieto *et al.*, 1989).

6.3.2 Micrococcaceae

Among Micrococcaceae, *Staphylococcus* as well as *Micrococcus* are used in commercial starter cultures (Hammes *et al.*, 1985). The effects desired from these genera are colour formation and colour stabilization together with aroma formation. *Staphylococcus* has become the most commonly used genus due to its ability to grow and metabolise under anaerobic conditions.

The nomenclature has changed in the course of time and some *Micrococcus* species have been renamed *Staphylococcus*. Among commercially available starter cultures the following species can be found: *Micrococcus varians*, *Staph. carnosus* and *Staph. xylosus*. Schleifer and Fischer (1982) describe *Staph. carnosus* as a new species isolated from dry sausages. Many of these strains were previously identified as *Staph. simulans*.

Colour formation.   For traditionally cured meats, nitrate is still used as curing agent. However, nitrate itself has no curing capability (Wirth, 1991) and reduction of nitrate into nitrite is necessary to obtain the desired effect. The reaction requires the presence of nitrate reductase, a bacterially produced enzyme. Among other microorganisms, Micrococcaceae possess this enzyme and the higher the activity of nitrate reductase the faster the colour formation.

Typical levels of nitrite in cured meat are 100–200 ppm although 30–50 ppm is enough to produce a correct cured colour (Wirth,. 1991). Leaving out nitrite will result in sausages with a greyish to a somewhat reddish appearance. The off-coloration depends on the degree of comminution and the casing calibre. Due to toxicological considerations, nitrite is currently the preferred curing agent. An initial nitrite level of about 100 ppm is reduced to 10–20 ppm after 10 days' fermentation and to less than 10 ppm after 30 days' sausage ripening (Wirth, 1991).

Also, if nitrite is used as sole curing agent, a high microbial nitrate reductase activity has been shown to accelerate the colour formation in European-type dried sausages. Nitrate is an end-product in the curing process and, therefore, a high nitrate reductase activity is highly favourable and results in a faster colour formation.

Nitrate reductase.   Nitrate reductase is an intracellular enzyme which is formed at the cytoplasmic membrane (Katsaras and Leistner, 1988). It promotes the dissimilation of nitrate at very low oxygen concentrations or under anaerobic conditions. Nitrate is used as electron acceptor and the microorganisms gain energy (Meisel, 1988).

It was shown by Puolanne *et al.* (1977) that the nitrate reductase activity of *Staph. carnosus* is almost unaffected by NaCl but optimum activity could be demonstrated at a pH value of 6.0 and at a temperature of 44°C. The activity is relatively stable within the pH range of 4.9–6.0. The enzyme is more exposed to changes in temperature, although at 20–22°C, the temperature range for European types of sausages, it still retains 50–60%

of the optimal activity. According to Meisel (1988), who examined two different strains of *M. varians*, the activity is species variable. Both strains were very much affected by the concentration of NaCl and temperature. They differed not only in pH optimum, one being optimal at pH 5.6 and the other at pH 6.5, but also in specific activity. Meisel (1988) stresses that evaluation of a potential starter strain cannot be based on the nitrate reductase activity alone. The strain needs to be metabolically active although propagation is not regarded as a necessity.

Colour stabilisation. The true cured colour arises from nitrosyl myoglobin which turns into nitrosyl myochromogen upon denaturation of the myoglobin. Both are sensitive to oxidation although nitrosyl myochromogen is far less so than nitrosyl myoglobin (Lücke, 1986a).

Microbially produced hydrogen peroxide is a well-known oxidative source in fermented meat (Lücke, 1986a). Hydrogen peroxide may lead to undesirable colour components in the meat, such as the green verdoheme and the yellow biliverdin (Slinde, 1984). Together with the red and brown pigments of the meat, the human eye will perceive this combination as grey spots or grey cores may even appear (Coretti, 1971). Both are typical faults associated with fermented sausages.

Hydrogen peroxide is decomposed in the presence of catalase, an activity common in Micrococcaceae (Lücke, 1986a).

Catalase. The most thoroughly described staphylococcal catalase is produced by *Staphylococcus aureus* MF-31. It has been reported that the synthesis of catalase, which is part of the cell defence system, is low during the initial exponential growth phase. The maximum level is reached at the onset of, or during, the stationary phase (Martin and Chaven, 1987). It has also been demonstrated that in sausage fermentations, catalase is formed late, during the ripening stage (Katsaras and Leistner, 1988).

The catalase activity is not induced by exogenous hydrogen peroxide (Galligan *et al.*, 1984). In the presence of glucose, it can be stimulated by exposure to the heme precursor 5-aminolevulinic acid, indicating that synthesis of the heme group is the rate-limiting step in the production of catalase activity (Galligan *et al.*, 1984). The authors also demonstrated that glucose alone results in a decreased enzyme activity. Glucose is catabolised fermentatively, thereby reducing the need for catalase as a defence against hydrogen peroxide.

A later publication on *Staph. aureus* MF-31 (Martin and Barrier, 1990) describes the influence of temperature, pH value and mineral salts. The enzyme is stable over a range of pH 4–9 and a temperature interval of 25–55°C. NaCl was more inhibitory than MgCl and KCl. NaCl concentrations of approximately 1 M reduced the catalase activity slightly. This level is somewhat higher than the inhibitory 3% NaCl reported by Galligan *et al.* (1984). The inhibitory effect was even more pronounced in combination with a low pH value, which is important in relation to sausage fermentation.

Aroma formation.   Several publications report the role of Micrococcaceae in the aromatisation of fermented meat products (Bacus, 1986, Lücke, 1986b). However, from literature it is difficult to deduce specific information on aroma components derived from the metabolic activity of Micrococcaceae which significantly influence the flavour of fermented meats.

Micrococcus varians, Staph. xylosus and *Staph. carnosus* all produce acid aerobically from glucose. Under anaerobic conditions *Staph. xylosus* is a somewhat weaker acid producer than *Staph. carnosus*. *Staph. carnosus* and *Staph. xylosus* produce acetoin anaerobically with the latter being a weaker producer (Kloos and Schleifer, 1986; Kocur, 1986).

Selected strains of *Staph. xylosus* which have been isolated from dry-cured ham were all able to decompose tributyrin whereas none was active against Tween 40 and Tween 80. A certain proteolytic activity was also demonstrated as out of 21 strains, five and three, were active against casein and gelatin, respectively (Carrascosa and Cornejo, 1991). This is in accordance with Nieto *et al.* (1989) who used subcutaneous ham tissue as medium and with results reported by Campanini *et al.* (1987) who found about 75% of the examined strains of *Staph. xylosus* to be lipolytic when tested on a pork fat medium. Nielsen and Kemner (1989) reported *Staph. carnosus* and *Staph. xylosus* to be lipolytic not only on tributyrin but also on animal fat from pork, beef and lamb meat.

Debevere *et al.* (1976) examined a *Micrococcus* sp. on pork fat. Free fatty acids were released followed by their breakdown and the subsequent formation of carbonyls. Some of these may have arisen because of bacterial activity.

6.3.3   Streptomyces

Streptomyces griseus sensu Hütter was introduced as a starter culture in 1977 with the purpose of obtaining a better aroma in addition to an improved colouring. Eilberg and Liepe (1977) describe the strain as being proteolytic, non-lipolytic, nitrate-reducing, catalase-positive and non-pathogenic.

Streptomyces griseus does not multiply during fermentation and exhibits a poor survival. In spite of this, a considerable decline in nitrate content was observed, resulting in a darker red colour. The flavour is described as aromatic and like old-fashioned cellar-ripened sausages (Eilberg and Liepe, 1977).

6.3.4   Yeasts

Dry-cured meat products such as sausages and ham are frequently contaminated with yeasts (Leistner and Bem, 1970), the yeasts being more typical on the surface than inside the product. The most common genera are *Debaryomyces, Candida* and *Torulopsis*. On Spanish dried ham, yeasts have been demonstrated in high numbers throughout the ripening period (Casado *et al.*, 1991).

In commercial starter-culture preparations, yeast is offered for exterior and interior use (Rudolf Müller; Rhône-Poulenc Texel; Laboratoires Roger). The strains used are all classified as *Debaryomyces hansenii*. The species is characterised by a high salt tolerance, no nitrate decomposition and high oxygen demands.

In the early 1970s, work on the use of yeast as starter culture for sausage fermentation was intensified (Coretti, 1972; Rossmanith *et al.*, 1972) and species of *Debaryomyces, Candida* and *Hansenula* were examined for sausage productions.

When added directly to the meat mix at an inoculation level of 10^6–10^7/g, the activity of the yeast is mainly observed in the periphery and the oxygen consumption accelerates the exterior colour formation. The addition of yeast results in a characteristic flavour particularly desirable for Italian types of sausages (Leistner and Bem, 1970; Rossmanith *et al.*, 1972; Coretti, 1977).

Coretti (1977) mentioned that the best results were obtained whenever the yeasts are used in combination with lactic acid bacteria and micrococci. This observation has recently been confirmed by Gehlen *et al.* (1991). These authors also stressed that the effect of the yeast is very much dependent on the presence or absence of other starters. Alone, yeasts result in an increase in ammonia content and a decrease in lactic acid and acetic acid. Taken together, the reactions result in a pH increase. The assimilation of lactic acid by *Debaryomyces hansenii* in sausages has been described by Ramihone *et al.* (1988). Gehlen *et al.* (1991) pointed out that because of the suppressive effect exerted by the yeast on the indigenous staphylococci, it is also necessary to use a strong nitrate-reducing organism. Preferably, lactic acid bacteria should also be included in order to obtain the optimal taste.

For surface treatment, yeasts can be used alone or in combination with a mould strain (Rhône-Poulenc Texel; Laboratoires Roger). The yeast is said to result in an ivory covering and adds an aromatic flavour to the sausages.

6.3.5   Moulds

Raw ham and to a certain extent fermented sausages are still being produced according to traditional methods, i.e. inoculation with the house flora. However, there is a big risk that toxinogenic strains are among those colonising the surface. According to Leistner (1984), it was demonstrated that out of 15 mycotoxins formed in a culture medium, 10 were also formed when the mould species grew on sausages or raw ham. Moulded ham is generally regarded as more hazardous than sausages because the casing acts as a kind of protector.

The genera most commonly encountered are *Penicillium* and *Aspergillus* but genera such as *Cladosporium*, *Scopulariopsis*, *Alternaria* and *Rhizopus* are also isolated (Leistner and Eckardt, 1979; Pestka, 1986). An Italian study of sausages (Grazia *et al.*, 1986) identified the most common and most dominant strain in small-scale manufacturing to be *Penicillium verrucosum* var. *cyclopium* and in industrial scale *Aspergillus candidus*. Both are known as highly toxinogenic species.

A Spanish study on dried ham revealed that *Penicillium* is frequently found throughout the ripening period whereas *Aspergillus* remains in low numbers until the ripening conditions change to a lower humidity and a higher temperature (Casado *et al.*, 1991). No other mould genera were identified in significant numbers.

In order to minimise health hazards by consumption of naturally moulded-meat products, the first non-toxic white mould culture was developed and made commercially available in 1972. It was a single-strain culture of *Penicillium nalgiovense* (Leistner, 1990).

Leistner and Eckardt (1979) have examined not only isolates from meat products but also moulded sausages from Italy, Hungary and Germany. From these, 77% of the meat isolates produced toxins. Some 77% of Hungarian isolates produced toxins whereas only 55% of Italian isolates produced toxins. In contrast, as little as 21% of the German isolates were positive, probably due to the use of the non-toxin producing strain *P. nalgiovense*.

The present-day mould strains used for surface treatment of dried sausages all belong to the genus *Penicillium*. The most frequently used one is *P. nalgiovense*, but *P. chrysogenum* and *P. camembertii* are also used (Lücke, 1986a; Lücke and Hechelmann, 1987). It is still not common to use mould cultures for dry-cured ham although in the future, legislation is expected to require well-characterised and non-toxic strains for ripening these products.

In addition to the white colouring and non-toxinogenic capacity, other desired qualities of a mould culture are the characteristic aroma and taste that it adds to the product. Positive side-effects such as reduced moisture loss due to mycelial covering, anti-oxidative effects because of catalase activity, oxygen consumption and reduced oxygen penetration through the mycelium are other factors mentioned among the advantages of using mould cultures (Bacus, 1986).

Mould cultures, commercially available in liquid or lyophilised form, are dissolved before use. They are added either by dipping the sausages into the solution or by spraying in the drying room. Use of mould cultures has been shown to suppress the natural moulds, thereby reducing the risk of mycotoxins (Leistner, 1986; Grazia et al., 1986). Depending on the ripening temperature, the moulds initiate the colonisation within a few days or a few weeks (Leistner, 1986). During ripening, the mycelium has been shown to penetrate the sausages completely, making it possible to influence taste and aroma even in the core. In addition to the lipolytic and proteolytic activities, many moulds also utilise the lactic acid. The typical pH rise noticed in mould-ripened sausages is mainly attributed to this utilisation rather than the ammonia production simultaneously occurring (Grazia et al., 1986). It should be noted that the sensory quality of moulded sausages is also influenced by the type of casing used. A deeper colonisation has been observed with natural casing in comparison to collagen casing. This was reflected in a superior sensory score with respect to intensity and quality of smell and flavour in the sausages fermented in natural casing (Roncalés et al., 1991).

Penicillium nalgiovense.   The popularity of this species is due to its growth ability on raw cured and dried meats together with its toxicological safety. However, it should be stressed that not all *P. nalgiovense* strains are non-toxic (Fink-Gremmels et al., 1988). Under the right environmental conditions, *P. nalgiovense* has not only proved its value on fermented sausages but also on dried ham (e.g. Südtiroler Bauernspeck) (Mintzlaff and Christ, 1973).

Penicillium chrysogenum.   According to Leistner (1990), *P. chrysogenum* is not suitable as a starter culture for sausage fermentation as the conidia become an undesirable greenish colour at temperatures above 15°C. In contradiction, Lücke and Hechelmann (1987) mention that a French manufacturer has selected and mutated a strain of *P. chrysogenum* and this produces a white covering. Compared to *P. nalgiovense* it grows more rapidly and has a higher optimum temperature which is favourable for sausage ripening.

   However, a disadvantage of *P. chrysogenum* is that practically all strains produce a mycotoxin or that in biological tests they turn out to be hazardous. Some 85% of 123 isolates have been shown to produce the mycotoxin roquefortine. In addition, use of the biological methods, which are indicative of acute toxicity (Fink-Gremmels et al., 1988), showed that for this reason 97% of the isolates were unsuitable as starter organisms for meat fermentation. Further, *P. chrysogenum* is well known for its penicillin production which may lead to allergic reactions in sensitive people (El-Banna et al., 1987).

Penicillium camembertii. Lücke (1986a) mentions that strains of *P. camembertii* are also beneficial to air-dried sausages regarding appearance and flavour. Unfortunately, *P. camembertii* has been shown to produce the mycotoxin cyclopiazonic acid at temperatures above 15°C (Still *et al.*, 1978). The production of cyclopiazonic acid is an almost universal property of *P. camembertii* strains as many laboratories have failed to isolate a strain incapable of producing this metabolite (Leistner and Eckardt, 1979; Scott, 1981).

6.4   Requirements for starter cultures

Starter cultures should be regarded as any other food additives, i.e. they should be GRAS-approved. The legislation varies from country to country but only a few countries demand official permits for the use of specific starter cultures. Among these is Denmark which requires a method of identification together with toxicological and virulence tests. In some countries, lists are submitted specifying genera or species permitted as starter cultures whereas other countries are without any regulation.

Requirements for starter cultures have been discussed in publications (Leistner *et al.*, 1979; Hammes, 1987) but no official definition exists. The following criteria should be regarded as fundamental for a standard:

Health aspect

• Only non-toxic and non-pathogenic strains should be used.
• The starter culture should be free from any microbial or chemical impurities that may cause health risks.

Technological performance

• The starter strain should possess activities that control the fermentation, i.e. control the acidification, affect flavour and colour development in a positive way and increase the microbial stability of the product.
• The starter strain should be phage-resistant in order to perform optimally during fermentation.
• The starter culture should not contain chemical or microbial components that inhibit or prevent manufacture.
• The starter culture should consist of strains that are phenotypically and genetically stable.

6.4.1 Lactic acid bacteria

Lactic acid bacteria have generally been regarded as safe organisms. However, in the 1970s a number of reports appeared on lactobacilli that had been implicated in human diseases involving septicaemia and abscesses (Kandler and Weiss, 1986). Among the potential pathogenic strains, *L. plantarum* has been mentioned. In contrast, pediococci have never been associated with diseases occurring in either animals or plants (Garvie, 1986).

According to Kandler and Weiss (1986), it has not yet been possible to describe the biochemical basis for pathogenicity. As a consequence, the authorities do not currently request pathogenicity tests on lactic acid bacteria.

Bacteriophage attacks on pediococci have never been described (Garvie, 1986) whereas lactobacilli phages have been reported (Kandler and Weiss, 1986). In sausages fermented with *L. plantarum*, a delayed pH drop was observed when Bradley's group B phages were co-inoculated (Nes and Sørheim, 1984) but not when Bradley's group A phages were used (Trevors *et al.*, 1984). Consequently, phage attacks may endanger an optimal fermentation process, although it was shown that the starter culture contained enough phage-resistant bacteria to ferment the sausage properly (Nes and Sørheim, 1984). A detailed description of bacteriophage B has been reported by Nes *et al.* (1988).

'6.4.2 Micrococcaceae

In contrast to the genus *Staphylococcus*, *Micrococcus* is considered to be non-pathogenic (Kocur, 1986). For this reason, staphylococci are still not approved in all countries as starter cultures for meat products. The word 'staphylococci' may give a negative impression because of the well-known toxicity of *Staph. aureus*. Also other species of staphylococci are known to cause various illnesses. Among the species used as meat starters, *Staph. carnosus* is regarded as non-pathogenic whereas *Staph. xylosus* and *Staph. simulans* have been reported to cause poisoning although they are said to be questionable pathogens (Kloos and Schleifer, 1986). Concerning *Staph. xylosus*, this may merely reflect inappropriate routines (Wadström and Rozgonyi, 1986). Strains among *Staph. xylosus* have been demonstrated to produce enterotoxin D (Bautista *et al.*, 1988).

Pathogenicity tests for staphylococci should comprise virulence determination and enterotoxin detection. Virulence determination can be carried out in mice. Cell suspensions and cell-free extracts should be injected into mice followed by biopsy. The enterotoxins A, B, C and D can be detected with the enzyme-linked immunosorbent assay (Bautista *et al.*, 1988).

Götz and Schleifer (1983) reported that *Staph. carnosus* was very sensitive to bacteriophages. This was later confirmed by Marchesini *et al.* (1991) who described phage attack on *Staph. carnosus* during sausage fermentation. The phages were species- but not strain-specific. Due to the limited growth of staphylococci during fermentation, the authors did not regard the phages as a serious problem but claimed, that in case of production faults, their possible existence should be taken into consideration.

6.4.3 Moulds

Penicillium species are known to be mycotoxin producers and every strain should be regarded as a potential health hazard. Prior to release of a strain as a starter culture, chemico-physical tests for mycotoxins should be examined together with biological tests for acute toxicity in a brine shrimp test and for dermato-toxic effects on rabbit skin (Fink-Gremmels *et al.*, 1988).

Penicillium is normally not regarded as allergenic. However, at high spore concentrations like those used in the sausage industry, the use of moulds has caused allergic alveolitis in humans (Filtenborg and Frisvad, 1988). *Penicillium nalgiovense*, *P. camembertii* and *P. chrysogenum* have all been shown to cause allergic reactions.

6.4.4 Quality control

According to Liepe (1983) the bacterial count and fermentation activity are directly correlated. Therefore, all starter cultures should be defined either by a cell number or by an activity test. Every starter-culture preparation should have a declared shelf life.

No official standardised method of analysis exists for meat-starter cultures. Most often, an acidification test is used for lactic acid bacteria and a nitrate-reductase activity test for the Micrococcaceae. However, the various suppliers mainly rely on internal analyses that cannot be compared because of differences in substrate and reaction kinetics. In addition, the absence of pathogenic microorganisms (*Staph. aureus, Clostridium, Listeria, Bacillus cereus*) and technologically undesirable bacteria (faecal streptococci, coliform bacteria and peroxide-forming bacteria) should always be controlled.

To ensure the stability of starter strains, the present analytical methods and instruments allow a phenotypic and a genetic characterisation. The former expresses metabolic activities that identify the specific strain. Genetic information can be obtained by means of DNA fingerprints and plasmid profiles. This allows monitoring of genetic stability over the years.

6.5 Production methods

Early starter cultures were sold as simple liquid cultures. Transportation problems prepared the way for more sophisticated production methods, resulting in today's frozen and freeze-dried concentrates. Lyophilised cultures are the most convenient for long-distance transportation. Frozen cultures are commonly shipped in styrofoam boxes packed with dry ice. Depending on the ambient temperatures, the cultures can remain frozen for 72–96 hours (Porubcan and Sellars, 1979).

A flow scheme for the large-scale production of starter culture concentrates is shown in Figure 6.1.

6.5.1 Cultivation

In order to obtain maximal biomass and enzyme activity the composition of the substrate and the fermentation conditions become a very critical point in the cultivation of starters. For the manufacturer, however, the choice of substrate is also a question of cost, efficiency in cell production, facility for harvesting, cell resistance to freezing and freeze-drying and stability during storage (Champagne et al., 1991).

The substrate nutrients can be classified in five groups consisting of nitrogen sources (yeast extract, peptones, protein hydrolysates etc.), carbon sources (simple or complex carbohydrates such as glucose, lactose or hydrolysed dextrins), growth factors (vitamins and mineral salts), buffers (organic or inorganic such as citrates, acetates and phosphates) and antifoamers (silicone or organic types) (Porubcan and Sellars, 1979; Champagne et al., 1991).

Important factors that influence growth and stability have been listed in a review by Champagne et al. (1991). Thus, suboptimal growth temperatures and addition of Tween 80 to the growth medium increase the proportion of unsaturated fatty acids in the cells which results in a better survival during freezing. The pH should be controlled, not only during cultivation but also in the suspension medium in order to obtain the highest survival rate during freeze-drying. Also the alkali used for pH adjustment is important and NH_4OH is preferred. Even if the pH is kept at optimal levels, biomass production is inhibited by the accumulation of lactate in the medium (Giraud et al., 1991). A lactate concentration of 110 g/litre completely inhibited a strain of L. plantarum.

Publications on cultivation of Micrococcaceae are few in number. However, lipase production from recombinant Staph. carnosus has been studied. The production of extracellular lipase was shown to be directly correlated with growth (Lechner et al., 1988). A low level of dissolved

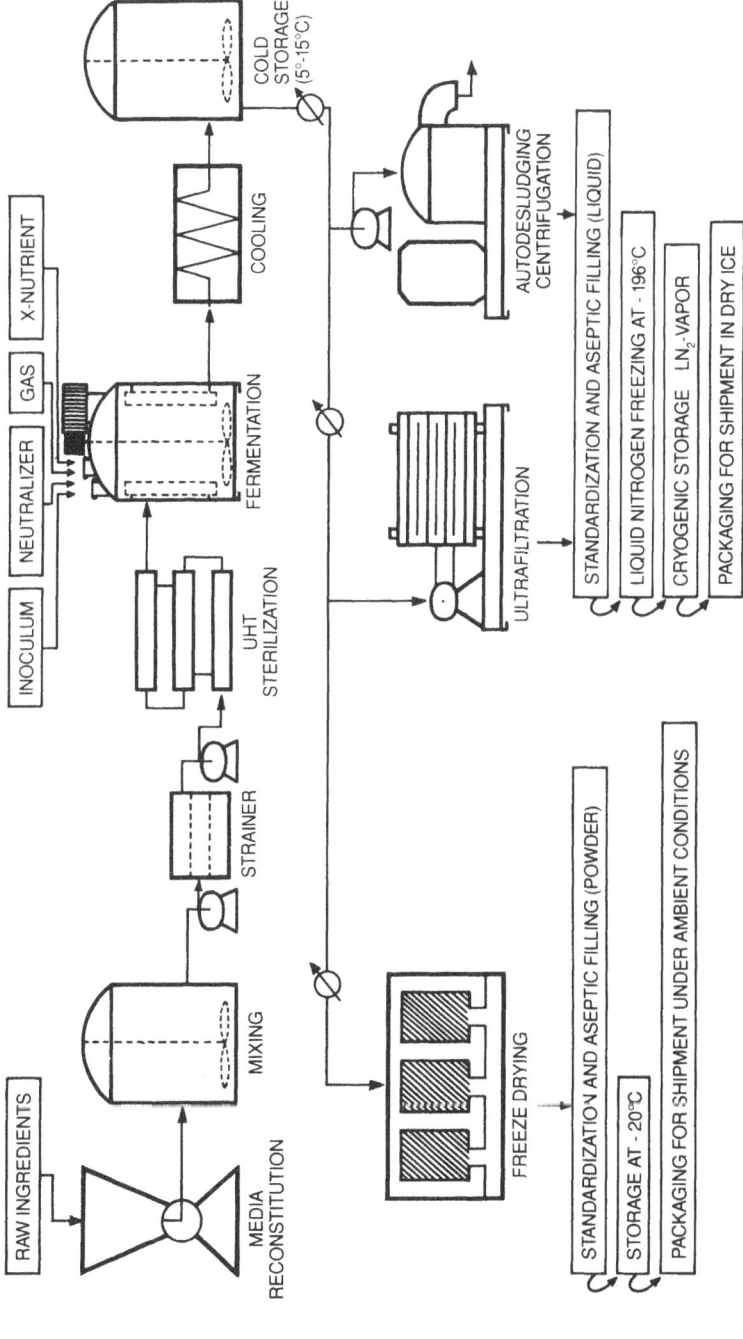

Figure 6.1 Flow scheme for the large-scale production of starter concentrates. Adapted with permission from Porubcan and Sellars (1979).

oxygen obtained by gentle stirring and low aeration rates was further shown to increase the lipase production.(Falk et al., 1991).

According to Porubcan and Sellars (1979) the sterile medium in the fermenter should be inoculated with 1–2% fresh culture. The cultivation is then commenced and parameters such as pH, temperature, agitation, oxygen pressure, nutrient concentration, etc. may be controlled automatically. At the end of the fermentation, usually during the late exponential or the early stationary phase, the culture is rapidly cooled before further processing.

6.5.2 Concentration and freeze-drying

Harvesting is the process of concentrating the fermentate by means of centrifugal separation or ultrafiltration. Depending on the nature of the bacterial culture and the machinery, a concentration factor of 20–100 can be obtained by centrifugation and 2–12 by ultrafiltration (Porubcan and Sellars, 1979). However, not all strains tolerate the high pressure exerted during centrifugation. Further, the sensitivity to freezing and drying may be increased due to stress caused by the centrifugation step (Champagne et al., 1991).

Concentrated microorganisms protect each other. This is due to a reduced interfacial area between the microorganisms and the external medium giving a shielding effect. However, the degree of concentration is important as an overconcentrated biomass, i.e. 10^{12}/ml, may be harmful because of the emergence of an unbalanced osmotic pressure (Bozoglu et al., 1987).

If the culture is to be sold frozen, the freezing can be carried out directly in the can immersed in the freezing medium. Liquid nitrogen is the optimal medium but other methods such as freezing in a blast freezer are also in use. The fastest and most advanced method is the freezing of droplets of starter-culture concentrate in an agitated bath of liquid nitrogen. The method results in pellets of 1–2 mm diameter that can be bulk-packaged as free-flowing particles. In order not to lose activity, the product should be kept at maximum of −45°C (Porubcan and Sellars, 1979).

Cultures for lyophilisation are first frozen and then dried under high vacuum, i.e. the water is removed by sublimation. The lyophilisation is done in steel cylinders rotating in the freezing medium, on steel trays in chambers equipped for injection of the freezing medium or by vacuum-drying of frozen pellets (Porubcan and Sellars, 1979).

Recent experiments have shown that immobilisation of the meat starters L. plantarum and P. pentosaceus prior to freeze-drying results in a 30% increase in acidification activity during meat fermentation (Kearney et al., 1990). Apparently, the additional nutrients in the immobilisation material promote the rapid activation of the cells upon rehydration.

After the lyophilisation, the powder is ground and packed according to a declared activity or cell number. Glucose or lactose are the carriers mostly used.

6.5.3   Protective compounds

Losses in viable cells are observed during freezing, freeze-drying and subsequent storage. Various factors such as change of cell-wall permeability, cold shock and metabolic injury are among those factors affecting survival (Bozoglu *et al.*, 1987). A certain compensation can be obtained by addition of cryoprotectants. Unfortunately, the protection afforded by a given additive appears to vary from strain to strain and further, a protection during freeze-drying does not automatically also guarantee protection during storage (Champagne *et al.*, 1991).

The following groups of substances have been tested for protective effects: polyols, polysaccharides, disaccharides, amino acids and proteins, vitamins, minerals and salts of organic acids (Champagne *et al.*, 1991). Glycerol is used with some success (Gilliland, 1977), and lactose plus saccharose are used successfully (Champagne *et al.*, 1991). The effect is partly attributed to their water-binding capacity and partly their ability to bind to the proteins, thus minimising the protein denaturation otherwise occurring. Glutamate and cysteine have a stabilising effect on the protein structure. Complex proteins such as peptones, yeast extracts, milk, gelatin and serum albumin have also been shown to give a protective effect (Champagne *et al.*, 1991). During storage, mixtures of carbohydrate and protein are useful with combinations of saccharose and gelatin being superior to skim milk (Rohde and Weckend, 1976).

6.5.4   Shelf life of starter cultures

The survival during storage of a starter culture, frozen or lyophilised, is a very important feature, not only to the manufacturer but also to the user. As already described, the stability of microorganisms is influenced by several parameters. In addition to those mentioned are temperature, time, water activity, gaseous atmosphere and method of rehydration.

High temperatures are detrimental, even to lyophilised microorganisms. It has been demonstrated that the lower the temperature, the better the survival of meat lactic acid bacteria (Rohde and Weckend, 1976; Liepe and Junker, 1981). Apparently, the surviving cells are the most resistant and also the most active ones as the cell loss is considerably higher than the loss in acidification activity during storage of lactobacilli for up to two years' storage (Liepe and Junker, 1981).

Tsvetkov and Brankova (1983) have examined the survival of *M. varians* and *L. plantarum*. They found an increasing cell loss during the storage period and a poorer survival in air compared to a vacuum. A thorough study of Bozoglu *et al.* (1987) on lactic acid bacteria confirms these results. The cell loss is most pronounced initially during storage. In this experiment, vacuum and also nitrogen, were superior to storage in air. The detrimental effect of air is probably due to oxygen that is assumed to diffuse into the freeze-dried cells.

According to Lücke (1986a), it is advisable to resuscitate lyophilised cultures several hours before use. However, this is in contradiction to Liepe and Junker (1985), who concluded that reactivation caused a very short increase in enzyme activity that was followed soon after by a considerable reduction in activity. The results were based on lactic acid bacteria and staphylococci, therefore the activities determined were acidification activity and nitrate reductase activity, respectively.

Depending on supplier, he declared shelf life of commercial lyophilised cultures ranges from three to seven months at 5°C and between one and two years at −18°C. Frozen cultures are given six months at −25°C whereas the pellet cultures have a shelf life of three months at −45°C. At higher temperatures, the pellets tend to stick to each other and an activity loss may be observed. However, this method guarantees that the frozen culture has not been thawed and refrozen which would have resulted in a decrease in cell count and activity.

Liquid mould cultures are given a shelf life from six weeks to six months at 0–5°C. The very long shelf life is allowed by dispersion in a hypertonic solution. Usually, a rapid dying out of liquid mould spores is observed upon exposure to high temperatures. Experiments have shown that no spores of *P. nalgiovense* survived four weeks at 25°C.

6.6 Strain improvement

Genetic manipulation of starter cultures can be used to improve strains, to incorporate desired traits or remove undesired qualities. The changes can occur in the chromosome and or on the plasmids. The methods adopted vary from simple mutagenesis through plasmid curing to advanced chromosomal transfers. Whenever genetic engineering is involved, it is important to ensure that the host organism and the vector are food grade and that the incorporated gene does not result in any pathogenicity or antibiotic formation (Hammes, 1988).

Many important characteristics are located on plasmids. This makes the characteristic unstable but facilitates curing work. Among the described characteristics that can be linked to plasmids are lactose fermentation, phage resistance, nisin production, proteolytic activity and citrate utilization (Barach, 1985).

Work on genetic modifications in starter cultures for meat fermentation has been initiated. A promising area seems to be the mould cultures. German authors (Leistner *et al.*, 1991a) describe that non-toxic strains of *P. camembertii* have been developed and also that lysostaphin genes have been transferred into *P. nalgiovense* giving strains which produce the component and inhibit growth of *Staph. aureus* during sausage fermentation. Furthermore, Leistner *et al.* (1991b) have isolated a DNA fragment that codes for a protease and have also been able to control glucose oxidase activity by co-transformation. Hereby, the possibility arises to control aroma formation and suppress rancidity.

In staphylococci, the main interest has been the antibiotic-resistance of *Staph. aureus*. It is believed that gene transfer in this respect between *Staph. aureus* and Gram-positive bacteria is possible and conjugation and phage-mediated conjugation are the most likely transfer systems to occur under natural conditions (Al-Masaudi *et al.*, 1991).

Plasmids are widely distributed in staphylococci. They can be transferred by the protoplast transformation method (Götz, 1990). By using this method it has been possible to express genes from other Gram-positive genera in *Staph. carnosus*. Due to the non-pathogenic nature of *Staph. carnosus*, Götz and Schleifer (1983) regard the species as very suitable for cloning experiments. However, it should be noted that the protoplast transformation is easier to carry out on some strains of *Staph. carnosus* than on others. In relation to meat fermentations, it is worth noting that the method has also been used to clone and express lipases from *Staph. hyicus*, *Staph. aureus* and *Staph. xylosus* in a non-lipolytic strain of *Staph. carnosus*. The details of substrate specificity of the cloned lipase from *Staph. hyicus* have been described by von Oort *et al.* (1989).

6.7 Future prospects

In the future, lactic acid bacteria may come to play a more important role as protective organisms in meat products including those hitherto not treated with starter cultures. The mode of action probably comprises several activities among which the production of bacteriocins looks promising.

Bacteriocins are proteinaceous substances with a bacteriocidal effect, most often against closely related species. Strains among *Lactobacillus* and *Pediococcus* have been demonstrated to possess an antimicrobial activity beyond formation of primary metabolites such as acids, hydrogen peroxide and ethanol. A very recent review on potential bacteriocins as meat preservatives lists the various possibilities (Stiles and Hastings, 1991).

It is characteristic that the bacteriocins are effective against Gram-positive bacteria among which *Listeria monocytogenes* and *Staph. aureus* are regarded as health hazards in meat products. However, several lactic acid bacteria, among which meat isolates can be found too, inhibited the Gram-negative *Aeromonas hydrophila* (Lewus *et al.*, 1991). Further, a strain of *P. acidilactici* isolated from fermented sausage produced an antimicrobial peptide designated AcH (Bhunia *et al.*, 1988). This bacteriocin has been shown to inhibit pathogens such as *Staph. aureus, Listeria monocytogenes* and *Clostridium perfringens* plus the specific spoilage bacteria *Brochothrix thermosphacta*. In addition, it acted against the two Gram-negative bacteria, *Aeromonas hydrophila* and *Pseudomonas putida*. Neither of these is important in respect of meat spoilage.

Besides the limitations in the spectrum of antimicrobial activity, the main disadvantages of most bacteriocins are the sensitivity to adipose tissue proteases, a significantly lower activity in meat systems than in broths and a certain pH decrease that is undesirable in non-fermented meats. Finally, there may also be legislative regulations to consider, as most authorities do not permit any kind of additive for fresh meat. However, recently nisin was approved by the Food & Drugs Administration as GRAS which may allow the use of other bacteriocins that are effective and well characterised (Delves-Broughton, 1990).

It is worth mentioning that a preservative effect based on a weak acid production by lactic acid bacteria is also possible (Jelle, 1989, 1991). Careful selection of appropriate lactic acid bacteria results in a suppression of the Gram-negative spoilage flora and the heterofermentative lactic acid bacteria and in addition, a specific inhibition of *Brochothrix thermosphacta* in vacuum-packaged meat is obtained.

Genetic manipulation has already been mentioned but without doubt, it is a subject that is very open for research and which is still in its infancy. The future will probably bring more sophisticated, more well-defined strains as regards flavour components that add the desired qualities to the product or even create new products.

Another, and not yet utilised, field in the meat industry is the manufacture and marketing of healthy meat products. Use of microorganisms such as *Lactobacillus acidophilus* and *Bifidobacterium bifidum*, perhaps in conjunction with low-fat, low-salt meat recipes, may open up a whole new product range in the meat industry.

References

Al-Masaudi, S.B., Day, M.J. and Russell, A.D. (1991) A review. Antimicrobial resistance and gene transfer in *Staphylococcus aureus. J. Appl. Bact.*, **70**, 279–290.

Bacus, J. (1984) *Utilization of Microorganisms in Meat Processing. A handbook for meat plant operators*. John Wiley, New York.

Bacus, J.N. (1986) Fermented meat and poultry products. In *Advances in Meat Research. Meat and Poultry Microbiology*. Eds A.M. Pearson and T.R. Dutson, pp. 123–164 AVI Publishing.

Bacus, J.N. and Brown, W.L. (1985) The lactobacilli: meat products. In *Bacterial Starter Cultures for Foods*. Ed. S.E. Gilliland, pp. 57–72 CRC Press, Boca Raton, FL.

Barach, J.T. (1985) What's news in genetic engineering of dairy starter cultures and dairy enzymes. *Food Technol.*, **39**(10), 73–84.

Bautista, L., Gaya, P., Medina, M. and Nuñez, M. (1988) A quantitative study of enterotoxin production by sheep milk staphylococci. *Appl. Environ. Microbiol.*, **54**(2), 566–569.

Berdagué, J.-L., Denoyer, C., Le Quéré, J.-L. and Semon, E. (1991) Volatile components of dry-cured ham. *J. Agric. Food Chem.*, **39**, 1257–1261.

Bhunia, A.K., Johnson, M.C. and Ray, B. (1988) Purification, characterization and antimicrobial spectrum of a bacteriocin produced by *Pediococcus acidilactici*. *J. Appl. Bact.*, **65**, 261–268.

Blickstad, E. and Molin, G. (1981) Growth and lactic acid production of *Pediococcus pentosaceus* at different gas environments, temperatures, pH values and nitrite concentrations. *Eur. J. Appl. Microbiol. Biotechnol.*, **13**, 170–174.

Bobillo, M. and Marshall, V.M. (1991) Effect of salt and culture aeration on lactate and acetate production by *Lactobacillus plantarum*. *Food Microbiol.*, **8**, 153–160.

Borch, E. and Molin, G. (1989) The aerobic growth and product formation of *Lactobacillus, Leuconostoc, Brochothrix* and *Carnobacterium* in batch cultures. *Appl. Microbiol. Biotechnol.*, **30**, 81–88.

Borch, E., Berg, H. and Holst, O. (1991) Heterolactic fermentation by a homo-fermentative *Lactobacillus* sp. during glucose limitation in anaerobic continuous culture with complete cell recycle. *J. Appl. Bact.*, **71**, 265–269.

Bozoglu, T.F., Özilgen, M. and Bakir, U. (1987) Survival kinetics of lactic acid starter cultures during and after freeze-drying. *Enzyme Microbiol. Technol.*, **9**, 531–537.

Campanini, M., Mutti, P. and Previdi, M.P. (1987) Caratterizzazione di Micrococcaceae da insaccati stagionati. *Industria Conserve*, **62**, 3–6.

Carrascosa, A.V. and Cornejo, I. (1991) Characterization of Micrococcaceae strains selected as potential starter cultures to Spanish dry-cured ham processes. 2. Slow process. *Fleischwirtschaft*, **71**(10), 1187–1188.

Casado, M.-J.M., Borrás, M.-A.D. and Aguilar, R.V. (1991) Fungal flora present on the surface of cured spanish ham. *Fleischwirtschaft*, **71**(11), 1300–1302.

Champagne, C.P., Gardner, N., Brochu, E. and Beaulieu, Y. (1991) The freeze-drying of lactic acid bacteria. A review. *Can. Inst. Sci. Technol. J.*, **24**(3/4), 118–128.

Chr. Hansen A/S data sheets, 10–12 Bøge Allé, DK-2970 Hørsholm, Denmark.

Condon, S. (1987) Responses of lactic acid bacteria to oxygen. *FEMS Microbiol. Rev.*, **46**, 269–280.

Coretti, K. (1965) Vorkommen und Bedeutung lipolytischer Mikroorganismen in Dauerwürsten. *Fleischwirtschaft*, **45**, 21–23.

Coretti, K. (1971) *Rohwurstreifung und Fehlerzeugnisse bei der Rohwurstherstellung*. Rheinhessischen Druckwerkstätte, Alzey, Germany.

Coretti, K. (1972) Kombinierte Anwendung von Milchsäurebakterien, Mikrokokken und Hefen als Starterkulturen bei der Rohwurst-Herstellung. *Jahresbericht der BAFF, Kulmbach*, **I**, pp. 48–49.

Coretti, K. (1977) Starterkulturen in der Fleischwirtschaft. *Fleischwirtschaft*, **57**, 386–394.

Debevere, J.M., Voets, J.P., De Schryver, F. and Huyghebaert, A. (1976) Lipolytic Activity of a *Micrococcus* sp. isolated from a starter culture in pork fat. *Lebensmittel-Wissenschaft Technol.*, **9**, 160–162.

Delves-Broughton, J. (1990) Nisin and its uses as a food preservative. *Food Technol.*, November, 100–117.

Eilberg, B.L. and Liepe, H.-U. (1977) Mögliche Verbesserungen der Rohwursttechnologie durch den Einsatz von Streptomyceten als Starterkultur. *Fleischwirtschaft*, **57**(9), 1678–1680.

El-Banna, A.A., Fink-Gremmels, J. and Leistner, L. (1987) Investigation of *Penicillium chrysogenum* isolates for their suitability as starter cultures. *Mycotox. Res.*, **3**, 77–83.

El Soda, M., Fathallah, S., Ezzat, N., Desmazeaud, M.J. and Abou Donia, S. (1986a) The esterolytic and lipolytic activities of lactobacilli. *Sciences des Aliments*, **6**, 545–557.

El Soda, M., Korayem, M. and Ezzat, N. (1986b) The esterolytic and lipolytic activities of lactobacilli. III. Detection and characterization of the lipase system. *Milchwissenschaft*, **41**(6), 353–355.

Everson, C.W., Danner, W.E. and Hammes, P.A. (1970) Bacterial starter cultures in sausage products. *J. Agric. Food Chem.*, **18**(4), 570–571.

Falk, M., Sanders, E. & Deckwer, W.-D. (1991) Studies on the production of lipase from recombinant *Staphylococcus carnosus*. *Appl. Microbiol. Biotechnol.*, **35**, 10–130.

Filtenborg, O. and Frisvad, J.C. (1988) Moulds and mycotoxins in Danish foods I [in Danish]. *Levnedsmiddelbladet*, January, 28–32.

Fink-Gremmels, J., El-Banna, A.A. and Leistner, L. (1988) Developing mould starter cultures for meat products. *Fleischwirtschaft*, **68**(10), 1292–1294.

Galligan, P.M., Barrier, W.A. and Martin, S.E. (1984) Factors affecting catalase activity in *Staphylococcus aureus* MF-31. *J. Food Sci.*, **49**, 1573–1576.

Garvie, E.I. (1986) *Pediococcus*. In *Bergey's Manual of Systematic Bacteriology*, Vol. 2. Eds P.H.A. Sneath, N.S. Mair, M.E. Sharpe and J.G. Holt, pp. 1075–1079 Williams & Wilkins, Baltimore, London, Los Angeles, Sydney.

Gehlen, K.H., Meisel, C., Fischer, A. and Hammes, W.P. (1991) Influence of the yeast *Debaryomyces hansenii* on dry sausage fermentation. *Proc. 37th Int. Congr. Meat Sci. Technol.*, *Kulmbach*, pp. 871–876.

Gilliland, S.E. (1977) Preparation and storage of concentrated cultures of lactic streptococci. *J. Dairy Sci.*, **60**(5), 805–809.

Giraud, E., Lelong, B. and Raimbault, M. (1991) Influence of pH and initial lactate concentration on the growth of *Lactobacillus plantarum*. *Appl. Microbiol. Biotechnol.*, **36**, 96–99.

Gottschalk, G. (1985) Bacterial metabolism, 2nd edn. Springer-Verlag, New York, Berlin, Heidelberg, Tokyo.

Götz, F. (1990) *Staphylococcus carnosus*: a new host organism for gene cloning and protein production. *J. Appl. Bact.*, Symp. Suppl., 49S–53S.

Götz, F. and Schleifer, K.H. (1983) *Staphylococcus carnosus* und seine Verwendung in der Gentechnologie. *Fleischwirtschaft*, **63**(11), 1758–1760.

Gray, J.I. and Pearson, A.M. (1984) Cured meat flavor. *Adv. Food Res.*, **29**, 1–86.

Grazia, L., Romano, P., Bagni, A., Roggiani, D. and Guglielmi, G. (1986) The role of moulds in the ripening process of salami. *Food Microbiol.*, **3**, 19–25.

Hammes, W.P. (1987) *Proceedings from Food Ingredients—European Conference on Ingredients and Additives, Wiesbaden*, 22 pp.

Hammes, W.P. (1988) Gefahren durch den Einsatz von Mikroorganismen in der Lebensmittelindustrie. *Alimenta*, **27**(3), 55–59.

Hammes, W.P., Rölz, I. and Bantleon, A. (1985) Mikrobiologische Untersuchung der auf dem deutschen Markt vorhandenen Starterkulturpräparate für die Rohwurstbereitung. *Fleischwirtschaft*, **65**(5/6), 629–636, 729–734.

Hammes, W.P., Bantleon, A. and Min, S. (1990) Lactic acid bacteria in meat fermentation. *FEMS Microbiol. Rev.*, **87**, 165–174.

Jelle, B. (1989) Biopreservation of vacuum-packed pork and poultry packaged in CO_2-atmosphere [in Danish]. Chr. Hansen Danmark A/S.

Jelle, B. (1991) Biopreservation of processed meat products [in Danish]. Chr. Hansen's Danmark A/S.

Kandler, O. (1983) Carbohydrate metabolism in lactic acid bacteria. *Antonie van Leeuwenhoek*, **49**, 209–224.

Kandler, O. and Weiss, N. (1986) *Lactobacillus*. In *Bergey's Manual of Systematic Bacteriology*, Vol. 2. Eds P.H.A. Sneath, N.S. Mair, M.E. Sharpe and J.G. Holt, pp. 1209–1234. Williams & Wilkins, Baltimore, London, Los Angeles, Sydney.

Katsaras, K. and Leistner, L. (1988) Topographie der Bakterien in der Rohwurst. *Fleischwirtschaft*, **68**(10), 1295–1298.

Kearney, L., Upton, M. and McLoughlin, A. (1990) Meat fermentations with immobilized lactic acid bacteria. *Appl. Microbiol. Biotechnol.*, **33**, 648–651.

Keller, H. (1954) Die bakterielle Aromatisierung der Rohwurst. *Fleischwirtschaft*, 1954(4), 125–126.

Klettner, P.-G. and Lücke, F.-K. (1992) Feinzerkleinerte streichfähige Rohwurst. Auswirkungen unterschiedlicher Starterkulturen. *Fleischwirtschaft*, **72**(3), 238–241.

Kloos, W.E. and Schleifer, K.H. (1986) *Staphylococcus*. In *Bergey's Manual of Systematic Bacteriology*, Vol. 2. Eds P.H.A. Sneath, N.S. Mair, M.E. Sharpe and J.G. Holt, pp. 1013–1035. Williams & Wilkins, Baltimore, London, Los Angeles, Sydney.

Kocur, M. (1986) *Micrococcus*. In *Bergey's Manual of Systematic Bacteriology*, Vol. 2. Eds P.H.A. Sneath, N.S. Mair, M.E. Sharpe and J.G. Holt, pp. 1004–1008. Williams & Wilkins, Baltimore, London, Los Angeles, Sydney.

Laboratoires Roger data sheets, B.P.20, F-77260 La Ferté sous Jouarre, France.

Lechner, M., Märkl, H. and Götz, F. (1988) Lipase production of *Staphylococcus carnosus* in a dialysis fermentor. *Appl. Microbiol. Biotechnol.*, **28**, 345–349.

Leistner, L. (1984) Toxinogenic penicillia occurring in feeds and foods: a review. *Food Technol. Aust.*, **36**(9), 404–406, 413.

Leistner, L. (1986) Mould-ripened foods. *Fleischwirtschaft*, **66**(9), 1385–1388.

Leistner, L. (1990) Mould-fermented foods: recent developments. *Food Biotechnol.*, **4**(1), 433–441.

Leistner, L. and Bem, Z. (1970) Vorkommen und Bedeutung von Hefen bei Pökelfleischwaren. *Fleischwirtschaft*, **50**(3), 350–351.

Leistner, L. and Eckardt, C. (1979) Vorkommen toxinogener Penicillien bei Fleischerzeugnissen. *Fleischwirtschaft*, **59**, 1892–1896.

Leistner, L., Linke, H., Eckardt, C., Lücke, F.-K. and Hechelmann, H. (1979) Anforderungen an Starterkulturen. Abschlussbericht zu einem Forschungsvorhaben. *Bundesanstalt Fleischforsch., Kulmbach*, 108 pp.

Leistner, L., Geisen, R. and Böckle, B. (1991a) Possibilities of and limits to genetic change in starter cultures and protective cultures. *Fleischwirtschaft*, **71**(6), 682–683.

Leistner, L., Fink-Gremmels, J., Geisen, R. and Böckle, B. (1991b) Optimierung von Schimmelpilz-gereiften Lebensmitteln. *Bundesministerium für Forsch. Technol, Bundesanstalt Fleischforsch. Kulmbach*, **10**, pp. 1–10

Lewus, C.B., Kaiser, A. and Montville, T.J. (1991) Inhibition of food-borne bacterial pathogens by bacteriocins from lactic acid bacteria isolated from meat. *Appl. Environ. Microbiol.*, **57**(6), 1683–1688.

Liepe, H.-U. (1983) Starter cultures in meat production. *Biotechnology*, **5**, 399–424.

Liepe, H.-U and Junker, M. (1981) Überlebensraten und Fermentaktivität von lyophilisierten Bakterienkulturen. *Arch. Lebensmittelhygiene*, **32**(5), 151–153.

Liepe, H.-U. and Junker, M. (1985) Untersuchungen zur Rehydratisierung (Reaktivierung) von Bakterien-Lyophilisaten. *Fleischwirtschaft*, **65**(3), 378–382.

Lücke, F.-K. (1986a) Fermented sausages. In *Microbiology of Fermented Foods*, Vol. 2. Ed. B.J.B. Wood, pp. 41–83 Elsevier Applied Science, London, New York.

Lücke, F.-K. (1986b) Microbiological processes in the manufacture of dry sausage and raw ham. *Fleischwirtschaft*, **66**(10), 1505–1509.

Lücke, F.-K. and Hechelmann, H. (1987) Starter cultures for dry sausages and raw ham. Composition and effect. *Fleischwirtschaft*, **67**(3), 307–314.

Lücke, F.-K., Popp, J. and Kreutzer, R. (1986) Bildung von Wasserstoffperoxid durch Laktobazillen aus Rohwurst und Brühwurstaufschnitt. *Chem. Mikrobiol. Technol. Lebensm.*, **10**, 78–81.

Martin, S.E. and Barrier, W.A. (1990) Influence of salt, pH and temperature on *Staphylococcus aureus* MF-31 catalase. *Food Microbiol.*, **7**, 121–127.

Martin, S.E. and Chaven, S. (1987) Synthesis of catalase in *Staphylococcus aureus* MF-31. *Appl. Environ. Microbiol.*, **53**(6), 1207–1209.

Marchesini, B., Bruttin, A., Moreton, R. and Sozzi, T. (1991) *Staphylococcus carnosus* bacteriophages isolated from salami factories in Germany and Italy. *Proc. 37th Int. Congr. Meat Sci. Technol., Kulmbach*, pp. 896–898.

Meisel, C. (1989) Mikrobiologische Aspekte der Entwicklung von nitratreduzierende Starterkulturen für die Herstellung von Rohwurst und Rohschinken. Ph D Thesis, University of Hohenheim, Germany.

Mintzlaff, H.-J. and Christ, W. (1973) *Penicillium nalgiovensis* als Starterkultur für 'Südtiroler Bauernspeck'. *Fleischwirtschaft*, 1973(6), 864–867.

Næs, H., Chrzanowska, J., Nissen-Meyer, J., Pedersen, B.O. and Blom, H. (1991)Fermentation of dry sausage — the imporance of proteolytic and lipolytic activities of lactic acid bacteria. *Proc. 37th Int. Congr. Meat Sci. Technol., Kulmbach*, pp. 914–917.

Nes, I.F. and Sørheim, O. (1984) The effect of infection of a bacteriophage in a starter culture during the production of salami dry sausage — A model study. *J. Food Sci.*, **49**, 337–340.

Nes, I.F., Brendehaug, J. and Husby, K.O. (1988) Characterization of the bacteriophage B2 of *Lactobacillus plantarum* ATCC 8014. *Biochimie*, **70**, 423–427.

Nielsen, H.-J.S. and Kemner, M.K.B. (1989) Lipolytic activity of meat starter cultures. *Proc. 35th Int. Congr. Meat Sci. Technol., Copenhagen, Denmark*, pp. 318–322.

Niinivaara, F.P. (1955) Über den Einfluss von Bakterienreinkulturen auf die Reifung und Umrötung der Rohwurst. *Fleischwirtschaft*, **35**, 603–605.

Niinivaara, F.P., Pohja, M.S. and Komulainen, S.E. (1964) Some aspects about using bacterial pure cultures in the manufacture of fermented sausages. *Food Technol.*, **18**, 147–153.

Nieto, P., Molina, I., Flores, J., Silla, M.H. and Bermell, S. (1989) Lipolytic activity of microorganisms isolated from dry-cured ham. *Proc. 35th Int. Congr. Meat Sci. Technol., Copenhagen*, Denmark, pp. 323–329.

Niven, C.F., Deibel, R.H. and Wilson, G.D. (1959) Production of fermented sausage. US Patent 2,907, 661.

Papon, M. and Talon, R. (1988) Factors affecting growth and lipase production by meat lactobacilli strains and *Brochothrix thermosphacta*. *J. Appl. Bact.*, **64**, 107–115.

Pederson, C.S. (1979) *Fermented Sausage. Microbiology of Food Fermentations*, 2nd edn, pp. 210–234. AVI Publishing, Westport, Westport, CT, USA.

Pestka, J.J. (1986) Fungi and mycotoxins in meats. In *Advances in Meat Research. Meat and Poultry Microbiology*, Vol. 2. Eds A.M. Pearson and T.R. Dutson, pp. 277–309. Avi Publishing, Westport, CT, USA, 1977(1).

Petäjä, E. (1977) Untersuchungen über die Verwendungsmöglichkeiten von Starterkulturen bei Brühwurst. *Fleischwirtschaft*, 1977(1), 109–112.

Porubcan, R.S. and Sellars, R.L. (1979) Lactic starter culture concentrates. In *Microbial Technology*, Vol. 1. Eds H.J. Peppler and D. Perlman, pp. 59–92. Academic Press, New York, San Francisco, London.

Puolanne, E., Törmä, P. and Djedjeva, G. (1977) Über das Nitratreduktionsvermögen der Stämme *Micrococcus* MIII und *Vibrio* 21. *Lebensmittel-Wissenschaft Technol.*, **10**, 7–11.

Raccach, M. (1985) Manganese and lactic acid bacteria. *J. Food Protect.*, **48**(10), 895–898.

Raccach, M. and Marshall, P.S. (1985) Effect of manganese ions on the fermentative activity of frozen-thawed lactobacilli. *J. Food Sci.*, **50**(3), 665–668.

Ramihone, M., Sirami, J., Larpent, J.P. and Girard, J.P. (1988) Gout acide des saucissons secs. *Viandes et Produits Carnes*, **9**(6), 291–298.

Reuter, G. (1975) Classification problems, ecology and some biochemical activities of lactobacilli of meat products. In *Lactic Acid Bacteria in Beverages and Foods*. Eds J.G. Carr, C.V. Cutting and G.C. Whiting pp., 221–229. Academic Press, London.

Rhône-Poulenc Texel data sheets, Z.A. de Buxières B.P.10, F-86220 Dangé-Saint-Romain, France.

Roca, M. and Incze, K. (1989) Starterkulturen für die Herstellung von Kochdauerwurst. *Fleischwirtschaft*, **69**(2), 175–181.

Rohde, R. and Weckend, B. (1976) Lyophil getrocknete Starterkulturen für die Fleischindustrie. *Fleisch*, **30**(3), 56–59.

Roncales, P., Aguilera, M., Beltran, J.A., Jaime, I. and Peiro, J.M. (1991) The effect of natural or artificial casing on the ripening and sensory quality of a mould-covered dry sausage. *Int. J. Food Sci. Technol.*, **26**, 83–89.

Rossmanith, E., Mintzlaff, H.-J., Streng, B., Christ, W. and Leistner, L. (1972) Hefen als Starterkulturen für Rohwürste. *Jahresbericht der BAFF, Kulmbach, I*, 47–48.

Rudolf Müller data sheets, Giessener Strasse 94, D-35415, Pohlheim, Germany.

Schleifer, K.H. and Fischer, U. (1982) Description of a new species of the genus *Staphylococcus: Staphylococcus carnosus. Int. J. Syst. Bacteriol.*, **32**(2), 153–156.

Scott, P.M. (1981) Toxins of *Penicillium* species used in cheese manufacture. *J. Food Protect*, **44**(9), 702–710.

Slinde, E. (1984) The colour of meat [in Norwegian]. In *Kjøttteknologi*, Ed. B. Underdal, pp. 71–78. Landbruksforlaget, Oslo.

Still, P., Eckardt, C. and Leistner, L. (1978) Bildung von Cyclopiazonsäure durch *Penicillium camembertii*-Isolate von Käse. *Fleischwirtschaft*, **58**, 876–877.

Stiles, M.E. and Hastings, J.W. (1991) Bacteriocin production by lactic acid bacteria: potential for use in meat preservation. *Trends Food Sci. Technol.*, October, 247–251.

Tetlow, A.L. and Hoover, D.G. (1988) Fermentation products from carbohydrate metabolism in *Pediococcus pentosaceus* PC39. *J. Food Protect.*, 51(10), 804–806.

Trevors, K.E., Holley, R.A. and Kempton, A.G. (1984) Effect of bacteriophage on the activity of lactic acid starter cultures used in the production of fermented sausage. *J. Food Sci.*, 49, 650–653.

Tsvetkov, T. and Brankova, R. (1983) Viability of micrococci and lactobacilli upon freezing and freeze-drying in the presence of different cryoprotectants. *Cryobiology*, 20, 318–323.

von Oort, M.G., Deveer, A.M.T.J., Dijkman, R., Tjeenk, M.L., Verheij, H.M., de Haas, G.H., Wenzig, E. and Götz, F. (1989) Purification and substrate specificity of *Staphylococcus hyicus* lipase. *Biochemistry*, 28(24), 9278–9285.

Vignolo, G.M., Ruiz Holgado, A.P. and Oliver, G. (1988) Acid production and proteolytic activity of *Lactobacillus* strains isolated from dry sausages. *J. Food Protect.*, 51(6), 481–484.

Wadström, T. and Rozgonyi, F. (1986) Virulence determinants of coagulase-negative staphylococci. In *Coagulase-negative Staphylococci* Eds. P.A. Måordh and K.H. Schleifer, pp. 123–130. Almquist & Wiksell, Stockholm.

Whittenbury, R. (1964) Hydrogen peroxide formation and catalase activity in the lactic acid bacteria. *J. Gen. Microbiol.*, 35, 13–26.

Wirth, F. (1991) Restricting and dispensing with curing agents in meat products. *Fleischwirtschaft*, 71(9), 1051–1054.

Wolf, G. and Hammes, W.P. (1988) Effect of hematin on the activities of nitrite reductase and catalase in lactobacilli. *Arch. Microbiol.*, 149, 220–224.

7 Stable and safe fermented sausages world-wide

L. LEISTNER

7.1 World-wide raw sausage technology

Raw sausage consists of raw, finely chopped meat and fat, which are mixed with salt, spices and a few additives, put into casings and, at the appropriate temperature and air humidity, fermented and dried for a sufficient length of time; the finished products are usually stored without refrigeration and consumed without being heated.

As far as Europe is concerned the raw sausage was apparently invented in Italy about 260 years ago and was thereafter adopted by other countries. Following a stay in Italy, a German butcher called Butleb is said to have started producing salami 210 years ago, and 155 years ago it is said that two Italian butchers initiated the production of the famous Hungarian salami in Budapest (Leistner, 1986b). Sausage preserved in its raw state (*lup cheong*) has been known in China for 2500 years. However, this product is not fermented and is only consumed in the heated state (Leistner and Dresel, 1986). At present, there is increasing world-wide interest in the European raw sausage, even in those countries, e.g. Australia and Brazil, where fermented meat products have received relatively little recognition.

Raw sausage production involves an imprecise technology, since, on the one hand, serious errors have to be made before a raw sausage becomes a faulty product, whilst, on the other hand, raw sausage technology allows for many variations, provided the basic concept is maintained (adequate reduction of pH and/or a_w). This is the basis of the wide variety of raw sausage products on offer (Leistner, 1986b). Attention here is drawn to some peculiarities of the raw sausage technology commonly used in different countries.

7.1.1 Raw material

For the top products in Italy, Hungary and France, pork is used exclusively for raw sausage, as the flavour and appearance of such products is preferred. In Germany a classical raw sausage recipe consists of one-third pork, one-third beef and one-third pork fat, with the addition of beef not being regarded as diminishing the quality. Pure beef sausages are rare amongst German raw sausages, nor are they easy to produce, as drying faults and, consequently, microbiological problems quickly occur. For religious reasons, Moslems, for example in Turkey, only make raw sausage from beef and the fat used comes from fat-tailed sheep. For this reason, these raw sausages (*Soudjouk*) have a specific 'sheep smell', which is popular in Turkey. The pork fat used for central European raw sausages is supposed to be robust and fresh. Under no circumstances may rancid fat ('freezer house fat') be made into raw sausage, as this leads to incorrect colour and faults in flavour, particularly in the case of slow-ripened products.

7.1.2   Additives

The addition of sufficient salt is essential for the raw sausage. According to the type, 2.5–3.0% is added and, in the case of spreadable sausage 2.8–3.0% can be expected in the finished product and 3.2–4.5% in sliceable sausage. The addition of nitrite (in the form of nitrite-curing salts) or nitrate is common in raw sausage, as fermented sausage produced with salt alone does not retain its colour well (grey centre) and quickly goes rancid. In many countries, such as the USA, nitrite and nitrate are used together in raw sausage. This is not allowed in Germany. Pure nitrate curing is also allowed for raw sausage in Germany, but only in products that are fermented for more than four weeks. The nitrite and nitrate residues are generally low in German raw sausages and are of the order of 10–30 ppm. The addition of sugar (carbohydrates) has been thoroughly dealt with by Wirth (1984). The addition of 0.3% glucose or saccharose to the raw sausage mixture is considered to be the optimum in Germany for slow-fermented products and 0.5–0.7% in the case of quick-fermented products. Lactose can be substituted for glucose or saccharose; in the case of slow-fermented products, 0.5% is recommended and for quick-fermented products 1.0% lactose. At fermentation temperatures of less that 15 °C, it is less important to add sugar. In Italy, the salami is generally fermented at low temperatures. The use of sugars in raw sausage in Italy was not permitted for a long time but now small amounts are regarded as useful. Spices are also important for raw sausage, particularly pepper, but also mace, cardamon and garlic are frequently used; no more than 1% of spices is usually added. Specialities, such as the Spanish *chourico*, may contain more than 2% paprika. The desirable fermentation bacteria of the raw sausage can be inhibited by large amounts of paprika, as well as by some spice extracts.

7.1.3   Chopping up

In German raw-sausage technology, the cutter has become the established means of chopping. Warming of the sausage mix is prevented by using well-chilled meat and frozen fat. In Italy and sometimes also in France they prefer to chop up the raw sausage mix in the grinder. This method produces not such a clear, even cut, although no great value is put on these criteria in the Romanic countries. The aroma, however, is important and to achieve this the raw sausage meat is prefermented for one to three days at 0–5 °C prior to putting it in the casing. In sliceable sausage, the degree of chopping depends on the type. In Germany the fineness increases from the *Plockwurst*, via *salami* through to *cervelat* ('dust fine'). With the similarly finely chopped spreadable raw sausage (*Mettwurst, Teewurst*) the binding of the sausage mix is not desirable and therefore the fat is first of all cut out until a lard-like consistency is obtained and the lean meat portion, together with the additives and spices, is mixed into the lard-like fat.

7.1.4   Filling

Raw sausage is put tightly into casings, that are permeable to water vapour, at a temperature close to 0 °C. Man-made casings (fibrous casings, collagen casings, cloth casings) or natural casings are available for this. The latter (e.g. fatty ends) are usually used for the top qualities; Milan salami manufacture preferably uses horse intestines. In France, 90% of the raw sausages produced are sold in natural casings. However, artificial casings are also used for high-quality products, e.g. fibrous casings for Hungarian salami. The diameter of the casings used can vary considerably; for spreadable raw sausages, 35-mm diameter artificial casings are generally used, whilst sliceable raw sausage is usually in skins of 65–90 mm diameter in Gemany. For mould-fermented raw sausage, a diameter of 30–40 mm should preferably be used, because only with a relatively small diameter in raw sausage will the mould aroma be fully developed (Leistner, 1986a).

7.1.5   Ripening

Even in the ripening conditions (temperature, relative air humidity, air velocity) there is a great variety of technologies that are used in the production of fermented sausage. In particular, the temperature and time of fermentation differ widely for the various products. It is generally true to say that the higher the temperature, the shorter the fermentation time. In the USA, for example, the half-dried 'summer sausage' is fermented for three days at 7 °C after putting in casings and then two to three days at 27–41 °C, then fermented again for two to three days at 10 °C. The product is then ready to be sold. However, if no frozen and, therefore, *Trichinella*-free pork is used, the meat must be heated again during the last four to eight hours of fermentation to an internal temperature of 58 °C. In Japan too, the finished raw sausage is heated to 68 °C, however, not because of the Trichinellae but in order to kill Enterobacteriaceae. In Italy, Milan salami is dried for 4–10 hours (according to the diameter) at 22 °C and 70–80% relative air humidity (RH) and then dried for a further three to four days at 20 °C and 80 to 90% RH. Thereafter the sausages are put into the fermenting rooms that are kept at no more than 15 °C and approximately 80% RH. A phased shifting of the relative humidity from 80 to 88% and again to 80% RH is regarded as beneficial for optimum drying. Fermenting takes up to six months, because only then the raw sausages will achieve their full aroma. In France, a similar technology to that in Italy is used for raw-sausage fermentation.

Unlike the Italian and French raw sausage, the raw sausages in Hungary ('winter salami') and Romania ('Hermannstädter salami') are lightly smoked for about two weeks at the start of fermentation, before mould colonisation. The smoke gives the products their characteristic flavour and also causes the sausages to become rancid more slowly. As in Italy, the winter salami in Hungary is kept at temperatures below 15 °C for most of the fermentation period, which takes 90–100 days. This also applies during smoking.

In China (as in other Asian countries) the process of making raw sausage (*lup cheong*) is quite different, using a technology that has clearly been developed independently of the European products. The products in question are coarsely chopped pork sausages. Stabilisation of this raw sausage is primarily achieved by a rapid reduction of the water activity (a_w), which, by virtue of drying at 48 °C and 65% RH is already below 0.92 after 12 hours and less than 0.90 in 36 hours (Leistner and Dresel, 1986). This rapid reduction in the a_w in Chinese raw sausage is made possible by the addition of sugar and salt, putting the product in pork casings of thin diameter (26–28 mm) and by drying over charcoal. Afterwards, the sausage is further ripened for three days at 20 °C and 75% RH, during which time the a_w drops below 0.80. The pH of *lup cheong* remains relatively high (about 5.9), for lactic acid bacteria should only achieve low counts because fermentation of the products and the accompanying acidic taste are undesirable. Raw sausages with high lactobacilli counts are even regarded as off because they taste too sour. What is desired is an aroma that is reminiscent of long-ripened raw ham ('good smelling sausage'), achieved by this technology. A further peculiarity of the Chinese raw sausage is the fact that this product is usually eaten hot, for example sliced in steamed vegetables (Leistner and Dresel, 1986).

7.1.6 Smoking or air-drying

In Germanic countries, raw sausages are usually smoked, whilst in Romanic countries this is only rarely the case. Thus, for example, in Western Germany, roughly 95% of the raw sausages are smoked and only 5% are mould-fermented, whilst the situation, for example in Italy, is exactly the reverse (Leistner, 1986a). By smoking, the aroma as well as the smell and taste of the raw sausages are influenced to the desired degree. Smoking can also prevent, or at least delay, any growth of bacteria (sausages become sticky) or yeasts and moulds. This is why in Germany the raw sausage is often lightly smoked even during fermentation. Of course, the amount of smoke must not be so much that the microbial fermentation processes inside the raw sausage (particularly close to the edge) are affected. The actual smoking of raw sausages is performed after fermentation has been completed and protection against moulds is also maintained whilst the sausages are stored. However, since the fungistatic substances in the smoke are volatile, even smoked products become mouldy after a while (particularly during dispatch). Therefore, in Western Germany, treatment of the finished raw sausages with potassium sorbate solution (20%) was recommended and approved. The non-smoked 'air-dried' raw sausages are usually, but not always, mould-fermented. Some have a desirable yeast coating ('sausage bloom') on the surface; others show absolutely no growth of microorganisms on the surface. In general, the inoculation of raw sausages with moulds should not use the 'house flora' of the fermentation rooms, as they will doubtless contain numerous toxinogenic moulds. For mould-fermented meat products, it is absolutely essential to use toxicologically safe starter cultures.

7.1.7 New developments

Over the last few decades the production of raw sausage has been optimised and standardised, particularly through the introduction of easily controlled fermentation chambers. However, this has led, at the same time, to increased energy consumption. Stiebing *et al.* (1982) suggested a much simplified method of controlling the atmospheric relative humidity in fermentation chambers, using fresh air. This produced a reduction in energy consumption for raw sausage fermentation of up to 70%. Further improvements appear possible, so that, for example, the climate in the fermentation chambers can be optimised and adjusted to the actual state of fermentation of a batch of raw sausage, by controlling the chamber by microprocessors via the surface a_w of the raw sausage. This can be determined by measuring the exact temperature beneath the surface of the sausage (Stiebing and Rödel, 1989).

7.2 Hurdles in fermented sausage

No microbiologist could have invented the raw sausage, because the process is in fact a monstrous one: raw meat and fat are stuffed into a casing,

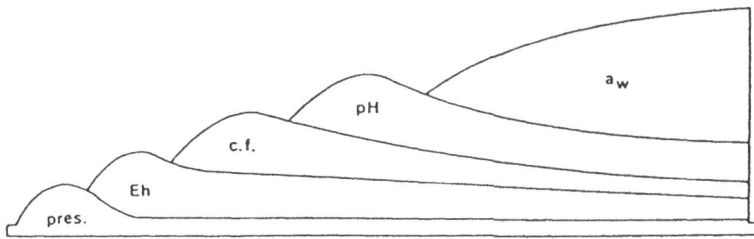

Figure 7.1 Sequence of hurdles in raw sausage. preso = preservation; Eh = redox potential; c.f. = competitive flora; pH = acidity; a_w = water activity.

where they are stored until consumed! The fact that this is at all practicable and the sausage does not become putrid is due, in the case of the European raw sausage, to the lactic acid bacteria and, particularly, to the lactobacilli. But the fact that these lactobacilli, which are originally usually only present in the raw material in relatively small counts, practically always assert themselves against the countless other microorganisms in a raw sausage mix, is attributable to the adding of salt and the exclusion of oxygen from the air, as well as the drying of the product. However, it is not only the lactobacilli that make the raw sausage into a meat product that keeps well; a series of factors have to interact, which can be explained as the 'hurdle effect' (Leistner, 1978). What hurdles can be expected in raw sausages and how do they influence the microbiological stability of a raw sausage? Figure 7.1 shows that there are several hurdles in a raw sausage which take effect one after the other in a specific sequence (Leistner, 1986b, 1987a).

The addition of the nitrite to the raw sausage mix, together with the nitrite-curing salt (far left in Figure 7.1) is particularly important at the start of fermentation for the microbiological stability of the product, since the other hurdles have not yet been established. The nitrite is important at the start of fermentation, especially for inhibiting any salmonellae that may be present in the raw sausage mix. At least 125 ppm nitrite should be added to the raw sausage, as is presently the case with the nitrite-curing salt used in Western Germany (Leistner, 1981). On the other hand, nitrate (saltpetre) encourages the growth of salmonellae in raw sausage (Hechelmann *et al.*, 1974). As with nitrate, the bacteriocidal nitrite is also broken down during the course of raw-sausage fermentation. Thus, the nitrite hurdle is short-lived and only effective in raw sausage at the beginning of fermentation.

The next important hurdle in the raw sausage during fermentation is the redox potential (E_h in Figure 7.1). During chopping, oxygen from the air is mixed in with the raw sausage batter. This results in a relatively high E_h. The addition of ascorbic acid or ascorbate and sugar reduces the E_h but more particularly the growth of bacteria which sets in at the start of raw-sausage fermentation, because many strains use up oxygen. This reduction in E_h acts on the raw sausage as a hurdle in different respects, i.e. in a positive way to stabilise the product; the nitrite is more effective as bacteriocidal substance when the E_h is low. Furthermore, aerophilic bacteria (particularly members of the family Pseudomonadaceae), which are often found in high counts in the sausage meat and which could result in the sausage going putrid, are inhibited by reduced E_h. What is especially important is that with reduced redox potential, the desirable lactic acid bacteria have a selective advantage, i.e. they can assert themselves over other microorganisms. In raw sausages with a slow fermentation (more than two months), the E_h tends to rise again, so that the redox potential hurdle becomes weaker. However, this is not a disadvantage, because in the meantime other hurdles have been built up in the raw sausage.

After the E_h the competitive flora (cf. in Figure 7.1) becomes the most important hurdle in a European raw sausage during fermentation. These are lactic acid bacteria which suppress undesirable microorganisms (e.g. Listeria, salmonellae, pathogenic staphylococci) in the raw sausage by lowering the pH and possibly also through the formation of bacteriocins. Lactic acid bacteria are of decisive importance in stabilising the European raw sausage. Since the bacteria count of the lactic acid bacteria falls again in slow-fermented raw sausages, the competitive flora is also only of rather short-lived significance in the stabilisation of the raw sausage.

The pH value (see Figure 7.1) is undoubtedly a very important hurdle in the stability of many raw sausages. Even the pH of the raw material (e.g. when processing more than 20% DFD meat) can make it difficult for the lactic acid bacteria to assert themselves. However, if the lactic acid bacteria do develop, as is normally the case, then the pH decreases. Particularly with quick-ripened products that still contain a lot of water and, therefore, have a relatively high a_w, the pH is an important hurdle. How quickly and how far the pH in a raw sausage drops can be affected by the amount of sugar added and the ripening temperature. The pH hurdle can also be enhanced by the addition of glucono-δ-lactone. During the further course of raw-sausage ripening, the pH tends to rise again, so that this hurdle becomes less important. In the case of raw sausage ripened slowly and at low temperature, like Italian salami, the pH does not go below 5.2, as is generally the case with the German raw sausage, and in fact does not fall below 5.4 at any time. The final pH is between 5.8 and 6.0. For stabilising products like this, the a_w is more important than the pH.

The a_w value (see Figure 7.1) of a raw sausage continues to fall as ripening progresses, so that the a_w is the only hurdle in a raw sausage which steadily increases in importance. How quickly and how far the a_w in a raw sausage falls is influenced by the recipe and the ripening temperature, but above all by the relative air humidity in the ripening chamber relative to the ripening time. The microbiological stability of slow-ripened raw sausage in particular is based on the a_w hurdle, where the pH is again raised and the residual nitrite content low. The bacteria count of the competitive flora has also dropped in slow-ripened products.

As a result of the sequence of hurdles described and illustrated in Figure 7.1, a raw sausage becomes a stable, safe product, because this sequence of hurdles inhibits not only the spoilage flora but also *Salmonella* spp., *Listeria monocytogenes*, *Staphylococcus aureus* and *Clostridium botulinum*. On the other hand, the desirable ripening organisms, i.e. particularly the lactic acid bacteria, are affected relatively little by the hurdles shown and they therefore have a selective advantage. The secret of the microbiology of the raw sausage is based on this sequence of hurdles.

7.3 Microbiology of fermented sausage

The microbiological processes that occur during the manufacture of fermented sausage have been clearly explained by Lücke (1985), and the raw sausage products that have failed as a result of microbial action have been described by Hechelmann (1985). The importance of food-poisoning bacteria for raw sausage has also been considered in respect of *Salmonella* spp. and *Listeria monocytogenes* (Schmidt, 1985, 1987, 1989; Leistner *et al.*, 1989) as well as *Staphylococcus aureus* and *Clostridium botulinum* (Katsaras *et al.*, 1985; Hechelmann *et al.*, 1988).

The most important microorganisms for raw-sausage ripening belong to the genera *Lactobacillus* and *Staphylococcus*; micrococci, yeasts and moulds are also significant. The following types of organisms are important for the 'normal' ripening of raw sausage: *Lactobacillus sake*, *L. curvatus* and *L. plantarum*; *Staphylococcus xylosus*, *Staph. carnosus* and *Staph. saprophyticus*; *Micrococcus varians*; *Debaryomyces hansenii*; *Penicillium nalgiovense*. Figure 7.2 shows, in diagrammatic form, the course of the a_w and pH, as well as the behaviour of the most important groups of organisms during 'normal' raw-sausage ripening (Lücke, 1985).

The lactobacilli form lactic acid and, thus, contribute to the typical taste of German raw sausage and help inhibit undesirable bacteria (salmonellae, *Listeria*, staphylococci). The Micrococcaceae (staphylococci and micrococci) reduce nitrate to nitrite and prevent or retard the development of discoloration and rancidity by forming catalase. At a higher E_h the Micrococcaceae tolerate a lower pH better and they can, therefore, be detected, particularly close to the edges of the raw sausage (Figure 7.2). The yeasts are similarly to be found mainly around the edges of the raw sausage; at the start of ripening, however, they can also be found in the centre of the products. Moulds generally grow only on the surface of the raw sausage; only in the case of products that have gone badly wrong,

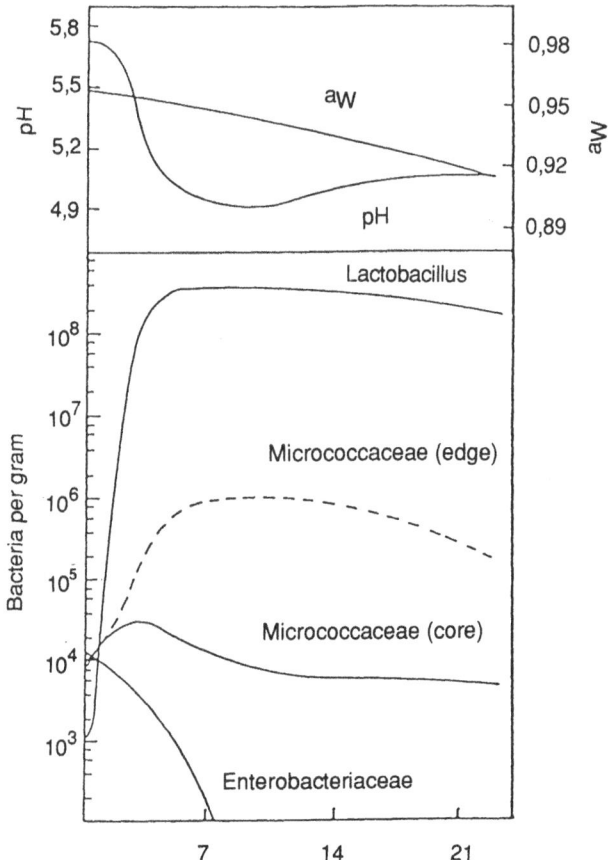

Figure 7.2 Diagrammatic representation of 'normal' ripening of fermented sausage (Lücke, 1985).

where severe cracking has occurred in the raw sausages, can moulds also be detected inside the products.

The current most important cause of raw sausage production going wrong is undesirable lactic acid bacteria, as these can cause overacidification, gas formation and discoloration. A sticky surface and greyish discolorations on the edges after washing the sausages are caused mainly by micrococci and yeasts. Moulds can cause a musty smell and taste as well as defective casings in raw sausages. Core putrefaction, which can be caused by Enterobacteriaceae (Hechelmann, 1985), is only observed relatively rarely.

Of the food-poisoning bacteria salmonellae and *Listeria* are currently of special interest in relation to fermented sausage. In Western Germany, salmonellae can be detected in roughly 4% of raw sausages. However, these only appear to result in food poisoning after there has been considerable growth (Schmidt, 1985). The use of 2.5% nitrite-curing salt, 0.3% glucono-δ-lactone and starter cultures, in addition to ripening temperatures below 25 °C, prevents the growth of salmonellae in raw sausage (Schmidt, 1987). *Listeria monocytogenes* is detectable in 15–60% of raw sausages, though usually with bacterial counts of less than 100 (Schmidt, 1989; Leistner *et al.*, 1989). The pH and the competitive flora are important in inhibiting the growth of *L. monocytogenes* in fermented sausage; if growth is not possible, then the *Listeria* count decreases with the ripening time. *Staphylococcus aureus* only presents a problem in German raw sausage at ripening temperatures above 25 °C (Katsaras *et al.*, 1985); however, in the case of mould-fermented raw sausage, *Staph. aureus* can grow even at lower temperatures because of the high pH (Hechelmann *et al.*, 1988). *Clostridium botulinum* is no problem for German raw sausage, for this food poisoner is incapable of growth even in products without added nitrite (Katsaras *et al.*, 1985). The growth of undesirable moulds on raw sausage must be strictly avoided, because most of the moulds that occur form mycotoxins. Apart from smoking, treating the sausages with potassium sorbate has proved effective in preventing the growth of undesirable mould. If toxicologically perfect starter cultures are used for mould-fermented raw sausages, then, once these cultures have started to grow, it is very difficult for the sausages to be colonised by undesirable moulds.

7.4 Topography of the bacteria in fermented sausage

Even at the start of ripening, the raw sausage mix contains numerous bacteria (10^5–10^6/g). These are, however, not evenly distributed. When the raw sausage changes from the sol state to the gel state during the course of ripening, the bacteria are fixed. Some types of organism are destroyed by the hurdles in the fermented sausage, others grow rapidly. This growth, however, can only take place in nests or small cavities, from which the bacteria cannot escape. If we look at these nests under the electron microscope, then we can see that they are situated far from one another (100–1500 μm) and the metabolic products formed in these nests, e.g. lactic acid, must consequently diffuse through an 'organism-free' matrix, in order to bring about ripening of the raw sausage and to inactivate undesirable bacteria, e.g. salmonellae (Katsaras and Leistner, 1988). Thus, raw-sausage ripening can be regarded as 'solid-state fermentation.' Through a better understanding of the topography of the bacteria in raw sausage, it is possible not only to control ripening, but the use of starter cultures may possibly be optimised (Leistner, 1987b).

7.5 Starter cultures for fermented sausage

It is also possible to produce excellent raw sausages without the use of starter cultures, because if appropriate ripening conditions are maintained, the desirable ripening bacteria that occur naturally in the raw material have a selective advantage and bring about the necessary preservation as well as the desired sensory properties (appearance, smell, taste, texture) of the fermented sausages. Even so, the use of starter cultures in raw sausage is on the increase and at present there are over 50 starter cultures preparations available in western Germany from roughly 10 producers (Hammes *et al.*, 1985). These starter cultures are used mainly with a view to the stability, standardisation and safety of production.

Although a patent was issued as early as 1940 to Jensen and Paddock for the use of lactobacilli as starter cultures for fermented sausage, the breakthrough in the USA in the use of starter cultures for raw sausage was only achieved as a result of the work of Niven *et al.* (1959) and Deibel *et al.* (1961a,b). In Europe the seminal work of Niinivaara (1955) and Nurmi (1966) helped the starter cultures to make their breakthrough. Starter cultures are particularly useful for quick-ripened fermented sausages and their use as protective cultures is generally on the increase in meat products and not only in raw sausage. A distinction between starter cultures, on the one hand, and protective cultures, on the other, has proved useful, with the starter cultures improving the sensory properties of food, whilst the protective cultures suppress undesirable microorganisms (e.g. salmonellae) in the food. Ideally, a starter culture should be a protective culture at the same time. Generally, one should be quite clear that the use of starter and/or protective cultures can only be one element of production control and quality assurance. Even starter cultures cannot offset gross production faults. Thus, to produce stable, safe and high-quality fermented sausage, perfect raw material and the effective control of temperature, relative humidity and air velocity are often more important than the use of starter cultures.

The main starter cultures used for fermented sausage are lactic acid bacteria, Micrococcaceae, yeasts and moulds. The bacteria are usually used in the form of compound preparations of lactic acid bacteria and Micrococcaceae to inoculate the raw sausage mix, to which yeasts also may be added. The moulds are used to inoculate the surface of the fermented sausages; occasionally yeasts are also used for this. Lücke and Hechelmann (1985) have discussed in detail the composition and effect of starter cultures for raw sausage. A concise summary follows.

7.5.1 Lactic acid bacteria

These microorganisms belong to several genera (*Lactobacillus*, *Streptococcus*, *Leuconostoc*, *Enterococcus*, *Pediococcus*); for raw sausage, primarily lactobacilli and pediococci and occasionally enterococci are used as starter cultures. The lactic acid bacteria result — particularly through sugar fermentation — in the formation of acid and therefore to a reduction in pH. This has a beneficial effect on the preservation, aroma, texture and drying of the raw sausage. When the pH is low, the breakdown of nitrite is encouraged and, thus, the colour of the raw sausage is improved. Through the formation of acid and antibiotic-like substances (bacteriocins) lactic acid bacteria act as protective cultures. Schillinger and Lücke (1989) isolated a *Lactobacillus sake* strain, which forms a plasmid-coded bacteriocin (Sakacin A), which is able to inhibit *Listeria monocytogenes* in microbiological media and, to a certain extent, also in meat products. The lactic acid bacteria used for fermented sausage are always homofermentative, i.e. they essentially form only lactic acid from glucose and other carbohydrates used in the production of raw sausages. Heterofermentative lactic acid bacteria would also form acetic acid and carbon dioxide, which are undesirable for raw sausage with respect to taste and appearance (porosity and cracks). Furthermore, lactic acid bacteria must not form peroxides which may lead to discoloration and rancidity in fermented sausages. The lactic acid bacteria used in raw sausage as starter and protective cultures belong mainly to the following species of organisms: *Lactobacillus plantarum*, *L. sake*, *L. curvatus*, *Pediococcus pentosaceus* and *P. acidilactici* (formerly *P. cerevisiae*).

7.5.2   Micrococcaceae

These bacteria are important for raw sausage in connection with the reduction of nitrate (colour formation in sausages cured with saltpetre and catalase formation (destruction of hydrogen peroxide). The aroma of raw sausage can be improved by the Micrococcaceae, which break down fat and protein and delay rancidity. Most of the starter cultures available for fermented sausage in Western Germany contain Micrococcaceae, particularly *Staphylococcus carnosus*, *Staph. xylosus* and *Micrococcus varians* (Hammes *et al.*, 1985). The technologically significant differences between these types of organisms are relatively slight (Lücke and Hechelmann, 1985). Micrococcaceae, which are added to the raw sausage meat as starter cultures, grow little or not at all during the ripening process (Reuter, 1972), therefore it is important that an adequate number of organisms is added.

7.5.3   Yeast

If yeasts are added to the raw sausage mix as starter culture (this is usually *Debaryomyces hansenii*), then they use up the oxygen and, by reducing the E_h, can cause the sausages to turn red rapidly. By breaking down fat and protein and forming specific metabolic products, yeasts can improve the aroma of the fermented sausage (Miteva *et al.*, 1986) and, by the formation of catalase, they can delay the onset of rancidity. In France, consumers like their raw sausages with a fine white coating ('sausage bloom') and so yeasts are also used there to inoculate the surface.

7.5.4 Moulds

Penicillium nalgiovense is used to inoculate the surface of mould-fermented raw sausages, since this species grows well on the raw sausage substrate. This mould not only gives the mould-fermented raw sausage its typical appearance, but it also protects the product from oxygen and light (delayed rancidity). A dense coating of a desirable mould on the surface also makes it difficult for undesirable moulds to become established. Furthermore, the breakdown of fat and protein and the formation of specific metabolic products affect the aroma of the raw sausages in a characteristic manner. For inoculation of raw sausage and raw ham, Mintzlaff and Leistner (1972) selected a non-toxic and technologically suitable strain of *Penicillium nalgiovense*, which is available commercially under the name 'Edelschimmel Kulmbach'. Strains of *P. nalgiovense* are also used in France for mould-fermented raw sausage. However, it must be noted that even the organism *P. nalgiovense* includes numerous strains, which form potent mycotoxins (Leistner, 1990). Thus, from a pool of 112 *P. nalgiovense* strains, Fink-Gremmels and Leistner (1990) were only able to isolate five strains, which are toxicologically and technologically suitable as starter cultures for mould-fermented meat products.

7.5.5 Requirements of starter cultures

Starter cultures must obviously not be pathogenic or toxinogenic, nor should they form any true antibiotics. Furthermore, they should not break down amino acids either into pharmacologically active amines (e.g. histamine) or into hydrogen sulphide. Over and above this, lactic acid bacteria should form little or no hydrogen peroxide, no gas and little or no acetic acid (Lücke and Hechelmann, 1985). The demands that have been made on starter cultures from a food law point of view have been discussed by Leistner *et al.* (1979).

The starter cultures used to date for fermented sausage are not necessarily the best suited strains. Too little is known as to what properties a starter or protective culture should or should not have. Therefore, attempts are presently being made to draw up requirement profiles for optimum strains, so that, from the natural reservoir, those strains can be selected, which best meet the requirements. If there are no optimum strains in the natural reservoir or if none is expected, then the starter or protective cultures best suited for a specific purpose could be 'tailor-made' using genetic methods. If only the mutation or conjugation is used, then the authorisation and acceptance of genetically modified strains should present no particular difficulties. Cultures modified by genetic engineering, which are transgenic microorganisms, do, however, require special authorisation and will only be accepted by the food consumer if they possess understandable advantages, e.g. in the way of health protection. Several laboratories are working on transgenic starter and protective cultures. Thus, for example, a *Penicillium nalgiovense* strain has been genetically engineered, into which the lysostaphin gene of *Staphylococcus staphylolyticus* has been transferred and which, consequently, appears to be well suited to mould-fermented raw sausage, because this strain brings about not only the desired sensory changes, but it can at the same time inhibit *Staphylococcus aureus* in the raw sausage (Geisen *et al.*, 1990). The use and therefore, the release of transgenic starter and protective cultures is not yet authorised and work with such cultures must, therefore, be classified as preliminary research.

7.6 HACCP concept for fermented sausage

Using the HACCP (Hazard Analysis Critical Control Point), the control points that are critical for production are derived from the analysis of the hygiene risks of a food. These control points can be monitored by guidelines and, therefore, neutralised (WHO, 1980). By strict application of the HACCP concept, stable and safe products can also be guaranteed in the case of fermented sausage. Leistner (1985) proposed 19 critical control points for raw sausage, taking into consideration the specific microbiological risks of this food and the important control points (raw material, sausage meat, ripening room, finished product) in raw-sausage production. A recommendation is given in respect of each control point, which, if monitored and adhered to, will prevent both food spoilage and food poisoning. Naturally, naming these 19 critical control points for raw sausage and the values that are vital for monitoring them cannot represent a final solution. Rather, these control points and the values assigned to them need to be subject of continuous discussion and examination. Also, the HACCP concept introduced by Leistner (1985) applies primarily to the products commonly available in Germany, whilst for other technologies (e.g. for Italian, Hungarian, US or Chinese raw sausages) appropriate modifications will have to be introduced. However, it is generally true that, in future, the HACCP concept will probably attract more attention, also in respect of the manufacture of meat products.

References

Deibel, R.H., Niven, C.F. Jr. and Wilson, G.D. (1961a) Microbiology of meat curing. III. Some microbiological and related technological aspects in the manufacture of fermented sausages. *Appl. Microbiol.*, **9**, 156–161.

Deibel, R.H., Wilson, G.D. and Niven, C.F. Jr. (1961b) Microbiology of meat curing. IV. A lyophilized *Pediococcus cerevisiae* starter culture for fermented sausages. *Appl. Microbiol.*, **9**, 239–243.

Fink-Gremmels, J. and Leistner, L. (1990) Toxicological evaluation of moulds. *Food Biotechnol.*, **4**, 579–584.

Geisen, R. Ständner, L. (1990) New mould starter cultures by genetic modification. *Food Biotechnol.*, **4**, 497–504.

Hammes, W.P., Rölz, I. and Bantleon, A. (1985) Mikrobiologische Untersuchungen der auf dem deutschen Markt vorhandenen Starterkulturpräparate für die Rohwurstbereitung. *Fleischwirtschaft*, **65**, 629–636, 729–734.

Hechelmann, H., Bem, Z. and Leistner, L. (1974) Mikrobiologie der Nitrat/Nitritminderung bei Rohwurst. *Mitteilungsbl. Bundesanstalt Fleischforsch.* No. 46, 2282–2286.

Hechelmann, H. (1985) Mikroblell verursachte Fehlfabrikate bel Rohwurst und Rohschinken. In *Mikrobiologie und Qualität von Rohwurst und Rohschinken*. Herausgegeben vom Institut für Mikrobiologie, Toxikologie und Histologie der Bundesanstalt für Fleischforschung, Kulmbach, pp. 103–127.

Hechelmann, H., Lücke, F.-K. und Schillinger, U. (1988) Ursachen und Vermeidung von *Staphylococcus aureus*-Intoxikationen nach Verzehr von Rohwurst und Rohschinken. *Mitteilungsbl. Bundesanstalt Fleischforsch.* N. 100, 7956–7964.

Jensen, L.B. and Paddock, L. (1940) Sausage treatment with lactobacilli. US patent No. 2,225,783.

Katsaras, K. and Leistner, L. (1988) Topographie der Bakterien in der Rohwurst. *Fleischwirtschaft*, **68**, 1295–1298.

Katsaras, K., Hechelmann, H. and Lücke, F.-K. (1985) *Staphylococcus aureus* und *Clostridium botulinum*. Bedeutung bel Rohwurst und Rohschinken. In *Mikrobiologie und Qualitat von Rohwurst und Rohschinken*. Herausgegeben vom Institut für Mikrobiologie, Toxikologie und Histologie der Bundesanstalt für Fleischforschung, Kulmbach, pp. 152–172.

Leistner, L. (1978) Hurdle effect and energy saving. In *Food Quality and Nutrition*. Ed. W.K. Downey, pp. 553–557. Applied Science, London.

Leistner, L. (1981) Neue Nitrit-Verordnung der Bundesrepublik Deutschland. *Fleischwirtschaft*, **61**, 338, 341–342, 344, 346.

Leistner, L. (1985) Empfehlungen für sichere Produkte. In *Mikrobiologie und Qualität von Rohwurst und Rohschinken*. Herausgegeben vom Institut für Mikrobiologie, Toxkologie und Histologie der Bundesanstalt für Fleischforschung, Kulmbach, pp. 219–244.

Leistner, L. (1986a) Schimmelpilz-gereifte Lebensmittel. *Fleischwirtschaft*, **66**, 168–173.

Leistner, L. (1986b) Allgemeines über Rohwurst. *Fleischwirtschaft*, **66**, 290–300.

Leistner, L. (1987a) Shelf-stable products and intermediate moisture foods based on meat. In: *Water Activity: Theory and Applications to Food* L.B. Rockland and L.R. Beuchat pp. 295–327 Marcel Dekker, New York.

Leistner, L. (1987b) Fermented meat products. *Proc. 33rd Int. Congr. Meat Sci. Technol.*, August 2–7, 1987, Helsinki, Finland, pp. 323–326.

Leistner, L. (1990) Mould-fermented foods: recent developments. *Food Biotechnol.*, **4**, 433–441.

Leistner., L., Linke, H. Eckardt, C., Lücke, F.-K. and Hechelmann, H. (1979) *Anforderungen an Starterkulturen für Lebensmittel tierischen Ursprungs*. Abschlussbericht zum Forschungsvorhaben des Bundesministers für Jugend, Familie und Gesundheit. pp. 108.

Leistner, L. and Dresel J. (1986) Die chinesische Rohwurst — eine andere Technologie. *Mitteilungsbl. Bundesanstalt Fleischforsch. No.* 92, 6919–6926.

Leistner, L., Schmidt, U. and Kaya, M. (1989) Bedeutung des Vorkommens von Listerien bei Fleisch und Fleischerzeugnissen. *Mitteilungsbl. Bundesanstalt Fleischforschung*, **28**, 440–445.

Lücke, F.-K. (1985) Mikrobiologische Vorgänge bei der Herstellung von Rohwurst und

Rohschinken. In: *Mikrobiologie und Qualität von Rohwurst und Rohschinken.* Her ausgegeben vom Institut für Mikrobiologie, Toxikologie und Histologie der Bundesanstalt für Fleischforschung, Kulmbach, pp. 85–102.

Lücke. F.-K. and Hechelmann, H. (1985) Starterkulturen für Rohwurst und Rohschinken, Zusammensetzung und Wirkung. In *Mikrobiologie und Qualität von Rohwurst und Rohschinken.* Herausgegeben vom Institut für Mikrobiologie, Toxikologie und Histologie der Bundesanstalt für Fleischforschung, Kulmbach, pp. 193–218.

Mintzlaff, H.-J. and Leistner, L. (1972) Untersuchungen zur Selektion eines technologisch geeigneten und toxikologisch unbedenklichen Schimmelpilz-Stammes für die Rohwurst-Herstellung. *Zentralbl. Vet. Med.* B, **19**, 291–300.

Miteva, E., Kirova, E., Gadjeva, D. and Radeva, M. (1986) Sensory aroma and taste profiles of raw-dried sausages manufactured with a lipolytically active yeast culture. *Nahrung*, **30**, 829–832.

Niinivaara, F.P. (1955) Über den Einfluss von Bakterien-Reinkulturen auf die Reifung und Umrötung der Rohwurst. Thesis, University of Helsinki; *Acta Agralia Fennica*, **85**, 1–128.

Niven, C.F. Jr., Deibel, R.H. and Wilson, G.D. (1959) Production of fermented sausage. US patent 2,907,661.

Nurmi, E. (1966) Effect of bacterial inoculation on characteristics and microbial flora of dry sausage. Thesis, University of Helsinki; *Acta Agralia Fennica*, **108**, 1–77.

Reuter, G. (1972) Versuche zur Rohwurstreifung mit Laktobazillen- und Mikrokokken-Starterkulturen. *Fleischwirtschaft*, **52**, 465–468, 471–473.

Schillinger, U. and Lücke, F.-K. (1989) Antibacterial activity of *Lactobacillus sake* isolated from meat. *Appl. Environ. Microbiol.*, **55**, 1901–1906.

Schmidt, U. (1985) Salmonellen, Bedeutung bei Rohwurst und Rohschinken. In *mikrobiologie und Qualität von Rohwurst und Rohschinken.* Herausgegeben vom Institut für Mikrobiologie, Toxikologie und Histologie der Bundesanstalt für Fleischforschung, Kulmbach, pp. 128–151.

Schmidt, U. (1987) Hemmung von Salmonellen durch technologische Massnahmen. *Mitteilungsbl. Bundesanstalt Fleischforsch.*, No. 96, pp. 7443–7448.

Schmidt, U. (1989) Verfahren zum Nachweis von Listerien in Fleisch und Fleischerzeugnissen. *Mitteilungsbl. Bundesanstalt Fleischforsch.*, **28**, 311–316.

Stiebing, A. and Rödel, W. (1989) Kontinuierliches Messen der Oberflächen-Wasseraktivität von Rohwurst. *Mitteilungsbl. Bundesanstalt Fleischforsch.*, **28** 221–227.

Stiebing, A. Rödel, W. and Klettner, P.-G. (1982) Energieeinsparung bei der Rohwurstreifung. *Fleischwirtschaft*, **62**, 1383–1389.

Wirth, F. (1984) Zur Wirkung von Zuckerstoffen bei Rohwurst. *Mitteilungsbl. Bundesanstalt Fleischforsch.*, No. 84, pp. 5924–5928.

World Health Organization (1980) Hazard analysis and critical control point. Application to food safety and quality assurance in developed and developing countries. Report on a meeting held June 9–10, 1980, at the WHO Headquarters in Geneva. 27 pp.

8 Flavour chemistry of fermented sausages

R. DAINTY and H. BLOM

8.1 Introduction

Flavour is a complex sensory phenomenon involving taste, smell (odour) and textural/tactile/mouthfeel responses (e.g. Forss, 1972) and flavour chemistry of foods has been, and continues to be, the subject of much research and speculation, not least in the case of the many different types of fermented sausages.

Fermented sausages have a flavour which to a greater or lesser extent is brand-specific and derived from the ingredients and chemical changes occurring during the fermentation and drying processes (ripening). Thirty years ago, Nurmi and Niinivaara (1964) stated: 'It is natural that the question of flavour formation in dry sausages is rather complicated for the reason that it is greatly affected, for example, by the smoking, by salt, nitrate and nitrite and their decomposition products, and by the spices, various bacterial ferments and the enzymes inherent in the meat.' They continued: 'Terplan (1962) [cited in Nurmi and Niinivaara, 1964] refers to the flavour formation caused by bacterial ferments and by the meat enzymes as fermentative flavour formation. In the first place it involves hydrolysis of proteins, fats, and carbohydrates.' With reference to numerous authors, they pointed out the importance attached at the time to fat hydrolysis, to the accumulation of carbonyl compounds and to the role of bacteria in producing chemical change. Since then, more qualitative and quantitative details of various aspects of these fat-degradation processes and the relative importance of microbial and endogenous meat enzymes have been obtained, together with similar information for carbohydrate and protein breakdown. Much of this activity occurred in the late 1960s and the 1970s, with particular emphasis on the northern European type of products. There followed something of a lull until, in the late 1980s and early 1990s, publications dealing with the same chemical phenomena in Spanish-type products began to appear. Additionally, analyses of headspace volatiles of fermented sausages using the highly sensitive and non-selective technique of capillary gas–liquid chromatography coupled to mass spectrometry have been published.

Despite this progress, it is fair to say that we are probably still a long way from understanding the relative importance of the various processes and products identified in flavour development. The result is that we are only able to catalogue what is known in chemical terms and to speculate on its significance.

The various compounds believed to play some role in flavour development will be described within the context of the four main processes traditionally regarded as being of greatest significance, i.e. glycolysis, proteolysis, lipolysis and lipid oxidation. It should be realised that the same end-products can be derived from more than one of the processes, so a certain amount of overlap is unavoidable. Finally we shall deal with headspace volatiles.

8.2 Glycolysis

The raw meat used in sausage making will contain up to 0.9 % by weight of L-lactic acid together with smaller amounts (<0.1 %) of glucose and phosphorylated glycolytic intermediates. The glucose and perhaps some of the phosphorylated intermediates, plus any carbohydrate added as an ingredient, are used by the indigenous flora of the meat, or starter cultures if used, as sources of energy for growth. The major consequence is the accumulation of more lactic acid primarily due to the fermentative metabolism of lactic acid bacteria, with production of both (L-) and (D-) isomers having been recorded (e.g. Lücke, 1986).

The relationship between lactate produced and hexoses utilised is illustrated in Figure 8.1 for a Belgian cold-smoked product (De Ketelaere, 1974). Clearly the amount of lactate produced, which was three times the initial level in the case illustrated, is related to the amount of carbohydrate present/added and therefore product dependent. Lactate may also be produced from amino acid metabolism, up to 10 % being so derived in the case of the type of Belgian product described above (Demeyer et al., 1986).

Acetic acid is another acid commonly detected in significant amounts in fermented sausages and believed to be of microbial origin (Table 8.1). Most of the products examined have had similar concentrations to those shown for the Belgian product in Figure 8.1. An exception was the *Cervelatwurst* examined by Langner (1969) which contained almost five times as much acetic acid, i.e. 10 mmol/100 g sausage. In the experimental sausages examined by Ordonez et al. (1989) clear evidence was obtained of a relationship between types of bacteria used to inoculate their salchichon-type sausage mix and the amounts of acetic acid produced. As in the case of lactic acid, hexoses and amino acids are believed to be a major source of acetate (Demeyer et al., 1986), but many lactic acid bacteria from sausages could also produce it along with lactate, from pentoses arising for example from ATP (Kandler, 1983; Lücke, 1986).

Whatever their sources, lactate and acetate are often suggested to be major contributors to acid aromas and tastes of fermented sausages. Such

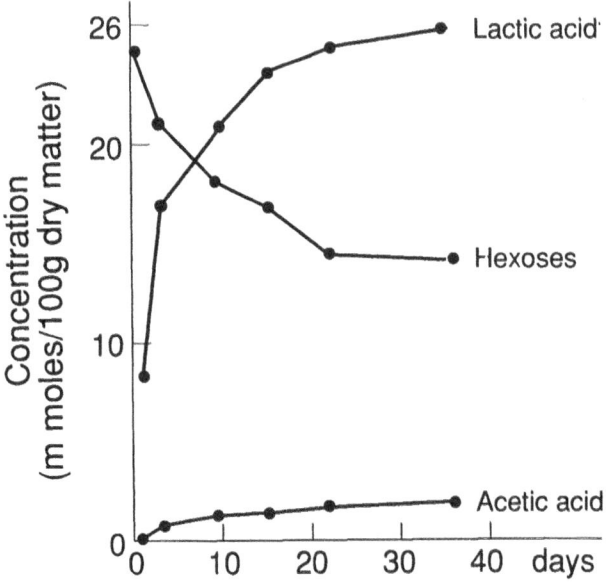

Figure 8.1 Changes in hexose, lactate and acetate concentrations during ripening of a Belgian cold-smoked sausage (from de Ketelaere *et al.*, 1974).

Table 8.1 Compilation from published data of short-chain organic acids detected in fermented sausages

Acid	Compound found in study[*]						
	1	2	3	4	5	6	7
Formic	×	×	×				
Acetic	×	×	×	×	×	×	×
Propanoic	×	×	×	×	×		
Butanoic	×	×	×	×	×	×	×
Pentanoic		×			×		
Hexanoic						×	×
2-Methylpropanoic	×	×			×		
3-Methylbutanoic	×				×	×	
2-Methylpentanoic						×	×
3-Methylpentanoic							×
Butenoic						×	
Propandioic						×	
Butandioic						×	
Methylenebutandioic						×	

[*] 1. Cantoni *et al.*, 1966. 2. Langner, 1969. 3. Halvarson, 1973. 4. De Ketelaere *et al.*, 1974. 5. Ordonez *et al.*, 1989. 6. Berger *et al.*, 1990. 7. Monteil, 1991.

flavours are perhaps more characteristic of northern European sausages, and particularly the rapidly fermented ones, than of the traditionally slower fermented southern European sausages (Langner, 1969; Demeyer *et al.*, 1986; Lücke, 1986). Although the association between acid flavours and the two acids seems a perfectly reasonable one, and the amounts of acetic acid are well above its determined flavour threshold detection levels (20–50 ppm) in aqueous and lipid systems (Forss, 1972), there appears to be little published evidence to directly prove this point. Too much acid production has been claimed to lead to undesirable 'sour', 'prickly', 'astringent' off-flavours (Lücke, 1986), while Demeyer *et al.* (1974) cites Burcharles as associating 'acidity' with excessive amounts of D-lactic acid.

In addition to these 'direct' effects lactate, through its major role in lowering pH (Demeyer *et al.*, 1979), and the effect this has on protein solubility (see below), can also be considered to play an important role in the development of the sausages' textural/tactile/mouthfeel properties which are another element in the overall appreciation of flavour (Forss, 1972; Dwivedi, 1975). Nor should the influence of pH on the ionisation of amino acid carboxyl groups (Demeyer *et al.*, 1979) and the effects this may have on aroma be forgotten.

Other acids have been reported in various combinations in a range of products (Table 8.1), their concentrations where measured being typically 100–1000-fold less than those of acetic acid. One exception to this was the finding of formic acid at concentrations of the same order of magnitude as those of acetic acid (only two to three fold less) in a Swedish fermented sausage (Halvarson, 1973). In view of the low odour and taste threshold levels (low ppm range) of many of these acids (Forss, 1972), the possibility that these compounds play a role in flavour development in some products cannot be discounted. The C_4 and C_5 acids have particularly characteristic odours and low thresholds. The source of these acids is unknown but the monocarboxylic acids with up to five carbon atoms are all well-established microbial metabolites formed by transamination or deamination to the corresponding α-keto acid, followed by decarboxylation of the latter (Dainty and Hibbard, 1983).

8.3 Proteolysis

The influence of pH on the extractability of meat proteins and the possible influence this may have on mouthfeel/texture of fermented sausages was mentioned above. In Figure 8.2, taken from De Ketelaere *et al.* (1974), substantial decreases in the proportions of total protein extractable as sarcoplasmic or myofibrillar protein using standard methods is evident. The extent to which extractability of these two fractions is lost during ripening has also been shown to vary from one product to another with one

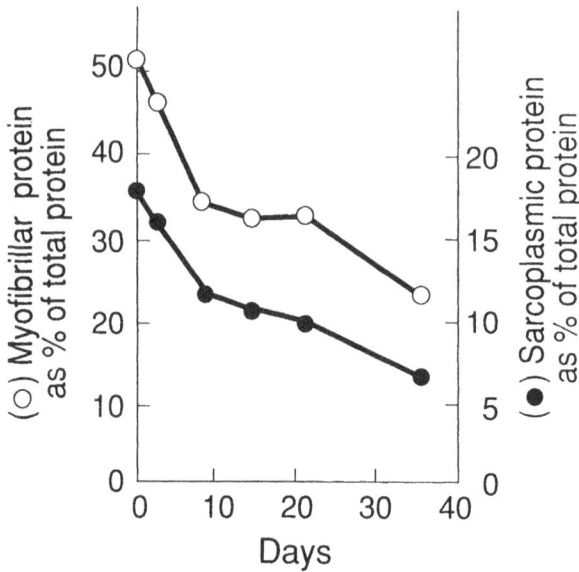

Figure 8.2 Changes in extractability of sarcoplasmic and myofibrillar protein fractions during ripening of a Belgian cold-smoked sausage (from De Ketelaere *et al.*, 1974).

fraction sometimes being more affected than the other (Pezacki and Pezacka, 1983; Astiasaran *et al.*, 1990; Garcia de Fernando and Fox, 1991).

On the other hand, a mass of evidence has been obtained over the years demonstrating the occurrence of proteolysis in a wide variety of sausage types. Reuter *et al.* (1968), Langner (1969), Sajber *et al.* (1971), Cantoni *et al.* (1974), Dierick *et al.* (1974), Lois *et al.* (1987), Astiasaran *et al.* (1990) and Garcia de Fernando and Fox (1991) have all provided confirmatory data based on one or more measurements of soluble non-protein nitrogen, α-amino nitrogen in peptides and/or amino acids, quantitation of ammonia or of individual amino acids, etc. The data in Figure 8.3 and Table 8.2 taken from Dierick *et al.* (1974) illustrates both quantitative and qualitative aspects of the phenomenon in a typical Belgian smoked product. Increases in α-amino nitrogen in free amino acids and peptides, ammonia and a range of individual amino acids are clearly evident. In view of the wide diversity of product formulations, temperatures and periods of ripening, smoked or air-dried etc., differences in the kinetics and the extent of the process are to be expected. Thus in the Belgian products described, faster production of free amino acids than of ammonia and peptides was observed, while Langner (1969) in his study of *Cervelatwurst* found ammonia to be produced earlier and faster than free amino acids. A perhaps more significant finding was the failure of Roncales *et al.* (1989) to detect any significant change in non-protein nitrogen in their study of

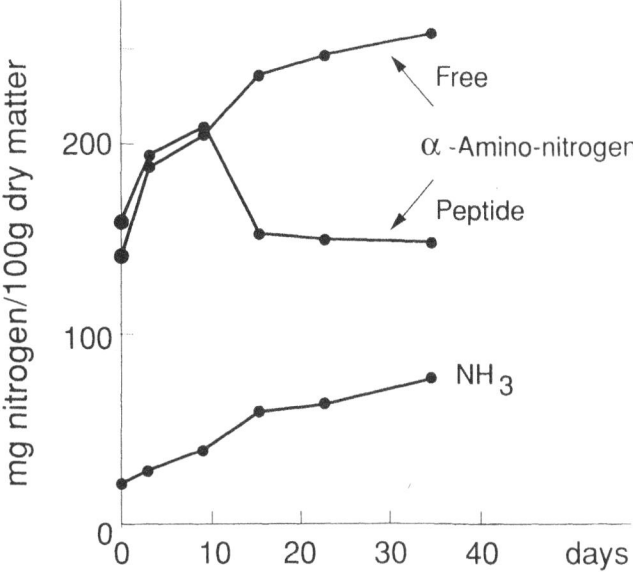

Figure 8.3 Changes in free and peptide-bound α-amino nitrogen and ammonia during ripening of a Belgian cold-smoked sausage (from Dierick *et al.*, 1974).

Table 8.2 Changes in concentration of free amino acids during ripening of a fermented sausage (from Dierick *et al.*, 1974)

Free amino acid	α-Amino nitrogen (mg/100g dry matter)	
	day 0	day 36
Aspartic acid	0.74	5.30
Threonine	0.82	25.20
Serine	1.73	9.10
Glutamic acid	19.00	5.60
Proline	0	5.50
Glycine	3.00	8.80
Alanine	10.20	25.20
Valine	1.44	8.85
Methionine	0.56	3.84
Isoleucine	1.60	5.45
Leucine	1.06	13.30
Phenylalanine	0.93	5.25
Lysine	2.07	6.35
Histidine	0.73	0.01
Tyrosine	0.77	0
γ-Aminobutyric acid	0	4.07
Ornithine	1.16	0

Spanish *fuet* sausages. Apart from these findings, there would appear to be a fair degree of uniformity between the various studies with total non-protein nitrogen values increasing up to two fold and individual amino acid concentrations by between two- and 12-fold (Table 8.2). Typically, the same amino acids, including valine, leucine, isoleucine, phenylalanine and methionine, are recorded amongst those showing the greatest increases in concentration.

In two studies (Verplaetse *et al.*, 1989; Garcia de Fernando and Fox, 1991) direct evidence of proteolysis has been obtained using sodium docecyl sulphate–polyacrylamide gel electrophoresis (SDS–PAGE). Data showing the breakdown of the myosin, actin and troponin T components of the myofibrils were obtained in both studies and the pattern of smaller molecular-weight peptides obtained was highly reminiscent of that produced by meat cathepsins. Verplaetse *et al.* (1989) pointed out that the pH (4.8–5.0) reached in their product and the temperature of ripening (15–20°C) were optimal for the activity of such enzymes and conjectured that the process could be regarded as a continuation of the meat-conditioning process. This is of interest because the question of the relative importance of microbial and non-microbial activities in proteolysis has been frequently addressed but must still be regarded as unresolved. Data from studies using different bacteria added to the initial sausage mix or the inclusion of antibiotics to suppress microbial growth point clearly to a role for microbes, with micrococci most frequently cited as having the necessary activity (Sajber, *et al.* 1971; Lücke and Hechelmann, 1986). However, from parallel increases in differential microbial counts and amino acid levels, lactobacilli have also been implicated (Reuter *et al.*, 1968) and De Masi *et al.*, (1990) have shown greater increases in non-protein nitrogen in sausages inoculated with pediococci than in non-fermented controls.

The proteolytic phenomena referred to above are quoted by many authors as being of importance in flavour development (e.g. Langner, 1969; Demeyer *et al.*, 1974; Verplaetse *et al.*, 1989). We have not been able to find any published confirmation of this conclusion and must assume it is based on the known sensory properties of amino acids and peptides and their substantial increases in concentration during ripening (Table 8.2). However, the majority of the amino acids stated above as showing the greatest concentration increases are recorded as having 'bitter' flavour characteristics (Dwivedi, 1975).

By virtue of its contribution to the increases in pH observed during drying of many types of sausage (Demeyer *et al.*, 1979) ammonia could also influence the sensory properties of compounds with ionisable groups. Ammonia production (Figure 8.3) and the implied further metabolism of amino acids is particularly evident in sausages with a long drying phase, as exemplified by the Spanish *Chorizo* sausages examined by Lois *et al.* (1987). Possible products of such metabolism have already been referred to, namely the short-chain volatile fatty acids detected in many products (Table 8.1).

The formation in various combinations of a range of amines including the so-called biologically active or pressor amines, i.e. histamine, tyramine and tryptamine, and putrescine, cadaverine, diaminopropane and 2-phenylethylamine is well recorded in fermented sausages (e.g. Vandek-erckhove, 1977; Tiecco *et al.*, 1985), Such compounds typically arise via the amino acid decarboxylating activity of bacteria, including some lactic acid bacteria from meat (Edwards *et al.*, 1987).

Formation of γ-aminobutyric acid from glutamic acid observed in certain products (Langner, 1972; Dierick *et al.*, 1974) presumably occurs by the same mechanism and may, according to the authors, render the use of glutamic acid as a flavour enhancer less effective than anticipated. Of these amines only putrescine (formed from arginine via ornithine) and cadaver-ine (formed from lysine) have any established sensory properties. How-ever, it is doubtful whether the amounts produced, typically up to a maximum of 300 μg/g sausage, are sufficient to influence taste or smell. This is probably fortunate in view of the unpleasant smells suggested by their names, and one goal of researchers working on the production of ideal starter strains by genetic techniques will be to rid the organisms of the ability to produce such compounds.

Monteil (1991) has provided evidence of the presence of the highly odoriferous dimethylamine in the headspace volatiles of experimental fermented sausages. No comment on its possible contribution to flavour was made.

8.4 Lipolysis

Release of long-chain fatty acids from neutral and phospholipids and from cholesterol has long been a central theme in investigations of chemical changes during sausage ripening (Nurmi and Niinivaara, 1964; Cantoni *et al.*, 1966; Cerise *et al.*, 1973; Demeyer *et al.*, 1974; Roncales *et al.*, 1989; Domínguez Fernández and Zumalacárregui Rodriguez, 1991). Increases in the levels of free fatty acids up to approximately 5% of the total fatty acid content have been typically found. An example of the time course of release is shown in Figure 8.4, the data again being taken from studies of Belgian sausages (Demeyer *et al.*, 1974). Free acid concentrations increased from the earliest sampling times with maximum levels being reached before the end of ripening. In other studies e.g. of Spanish *chorizo* (Lois *et al.*, 1987), levels were still increasing at the final sampling time.

Demeyer *et al.* (1974) further showed that as the amount of free acids increased the amount esterified as triglyceride decreased while the amount in di-, and to a lesser degree monoglyceride form, increased (Figure 8.4). Over the same period, there was no measurable change in phospholipid

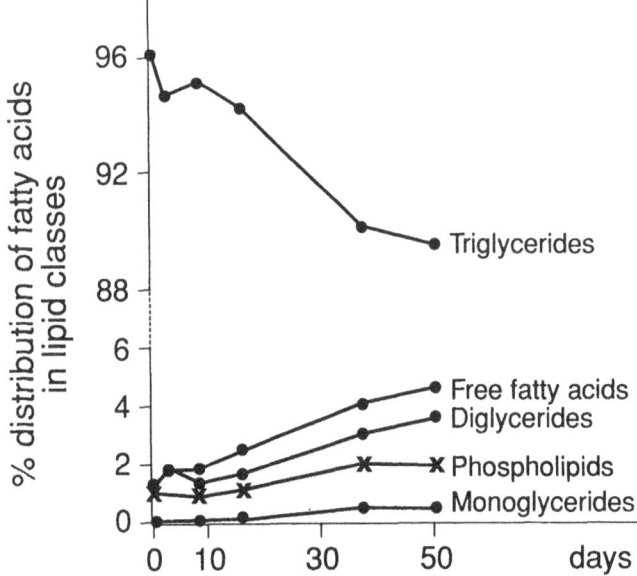

Figure 8.4 Changes in distribution of free acids in lipid classes during ripening of a Belgian cold-smoked sausage (from Demeyer *et al.*, 1974).

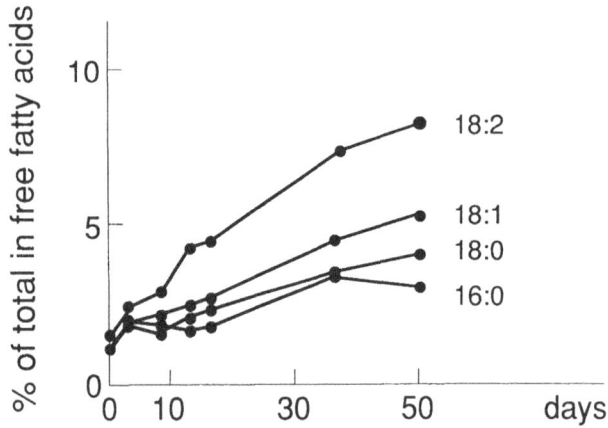

Figure 8.5 Time course of appearance of saturated and unsaturated fatty acids in a Belgian cold-smoked sausage (from Demeyer *et al.*, 1974).

fatty acid content though the authors stated that longer storage could leadto the release of acids from this source as well. Such a phenomenon was recorded by Cantoni *et al.* (1966) who also detected release of acids bound to cholesterol. The data in Figure 8.5 taken from Demeyer *et al.* (1974) show that the rate of release of individual acids decreased in the order linoleic > oleic > stearic > palmitic. In view of the preponderance of unsaturated acids in position 3 in triglycerides of porcine fat, the results were taken to indicate enzyme specificity, a characteristic of bacterial lipolysis previously recorded by Alford (1971). Similar findings in sausages had been reported earlier by several groups of workers (Cantoni et al., 1966; Wahlroos and Niinivaara, 1969; Cerise et al., 1973). However, no evidence of such specificity could be found in a Spanish study (Domínguez Fernández and Zumalacárregui Rodriguez, 1991). Regarding the agent(s) of lipolysis, there appears to be a general consensus from the use of different bacterial cultures as inocula (Wahlroos and Niinivaara, 1969) and from the coincidence of increases in free fatty acids and numbers of lipolytic bacteria that the latter play an important role. Micrococci are thought to be particularly active as are moulds and yeasts in certain products (Lücke and Hechelmann, 1986). A role for meat enzymes cannot be ruled out however (Dobbertin et al., 1975).

There are no comments in any of the studies referred to regarding possible sensory consequences of the accumulation of free fatty acids and indeed oleic and linoleic acids are said to be odourless (Forss, 1972). It is perhaps reasonable to suggest that the carboxyl groups of the de-esterified acids might add to the 'acid', 'sour' tastes of the products. What is generally considered to be of greatest significance for flavour, however, is the role these acids, and the unsaturated ones in particular play as precursors of smaller molecular-weight compounds with highly characteristic flavour properties.

8.5 Lipid oxidation

Of the authors quoted in the previous section only Cerise et al. (1973) in their study of pure pork salami provided direct evidence of decreases in the levels of certain long-chain acids, including oleic acid, during the later stages of ripening after earlier increases in concentration. They attributed the decreases to oxidation reactions leading in part to the accumulation of acids of shorter chain length (up to C_{10}). They specifically identified a C_{12} acid as α-butenyl-β-ketooctanoic acid and a C_{11} compound as α,α-dimethyl- or α-ethyl-undecanoic acid. They further found that the decrease in oleic acid concentration coincided with a proportional increase in a range of other acids having carbonyl groups though these were not identified further. A series of aliphatic straight-chain compounds (C_9–C_{20}) and related 2-methyl derivatives (C_8–C_{13}) with unrecorded functional groups were also found, together with a C_{16} cyclic ketone, dodecene, tetradecene, tridecadiene and tetradecadiene. These types of compound are well established end-products of autoxidation of unsaturated fatty acids (Mottram, 1991). In this context, Cerise et al. (1973) detected a rapid build up of peroxides, the first end-products of autoxidative reactions, during the

Table 8.3 Compilation from published data of aldehydes detected in fermented sausages

Aldehyde	Compound found in study*						
	1	2	3	4	5	6	7
Methanal		×	×	×			
Ethanal	×	×	×	×		×	
Propanal	×	×	×	×			
Butanal	×	×	×				
Pentanal	×		×	×		×	×
Hexanal		×	×	×	×	×	×
Heptanal		×	×	×	×	×	
Octanal		×	×	×	×		
Nonanal			×		×	×	×
Decanal			×	×			
2-Butenal		×					
2-Pentenal			×				
2-Hexenal		×	×	×		×	
2-Heptenal			×	×			×
2-Octenal			×				×
2-Decenal			×				
2-Undecenal			×				
2,4-Hexadienal				×			
2,4-Heptadienal				×			
2,4-Octadienal						×	
2-Methylpropanal	×	×	×	×		×	
2-Methylbutanal				×		×	
3-Methylbutanal	×			×		×	×
2-Methyl-2-butenal					×		
2-Methyl-2-pentenal					×		
Methylpentanal			×				
2-Pentenal (branched)				×			
2-Hexenal (branched)				×			

* 1. Cantoni *et al.*, 1966. 2. Langner, 1969. 3. Langner, 1972. 4. Halvarson, 1973. 5. Berger *et al.*, 1990. 6. Monteil, 1991. 7. Edwards *et al.*, 1991.

first days of ripening when oleic acid levels were also increasing. There was a subsequent rapid and extensive fall in the peroxides while oleic acid levels declined more slowly. More evidence interpreted as indicative of oxidative changes to lipid components during ripening has been obtained using methods such as peroxide values, TBA numbers, benzidine numbers for aldehydes, and spectrophotometric and chromatographic determination of total and individual carbonyls (Cantoni *et al.*, 1966; Wahlroos and Niinivaara, 1969; Langner, 1969, 1972; Halvarson, 1973; Demeyer *et al.*, 1974; De Ketelaere *et al.*, 1974; Nagy *et al.*, 1989). A compilation of the compounds identified by some of these authors is given in Tables 8.3 and 8.4.

Of the compounds recorded by Cantoni *et al.* (1966), methanal (acetaldehyde), pentanal, 2-butanone, 2-pentanone and 2,3-butanedione (diacetyl) increased in concentration but 3-methylbutanal decreased. Langner (1969) found increases in 2-hexenal, methanal, ethanal, 2-propanone (acetone) and 2,3-butanedione and commented on their generally held importance by referring to the use of a carbonyl index in quality control. In

Table 8.4 Compilation from published data of ketones detected in fermented sausages

Ketone	Compound found in study*						
	1	2	3	4	5	6	7
Propan-2-one	×	×	×	×		×	
Butan-2-one	×	×	×	×		×	
Pentan-2-one			×	×	×	×	
Hexan-2-one			×			×	
Heptan-2-one			×		×	×	×
Octan-2-one			×		×	×	×
Nonan-2-one			×		×		×
Decan-2-one			×				
Undecan-2-one			×		×		
Tridecan-2-one			×				
Pentdecan-2-one			×				
Pentan-2-one	×			×	×		
Octan-3-one							×
4-Methylpentan-2-one					×		×
3-Methylhexan-2-one						×	
5-Methylhexan-2-one					×		
6-Methyl-5-hepten-2-one						×	×
Dimethyloctenone					×		
Butan-2,3-dione	×	×		×		×	×
Pentan-2,3-dione							×
Octan-2,3-dione						×	
3-Hydroxybutan-2-one						×	×

* See Table 8.3, footnote.

a later study of a range of commercial products including mainly salami and *Cervelatwurst*, the same author (Langner, 1972) recorded a much greater number of carbonyls and showed that sausages with the greatest aroma had the highest levels of carbonyls with between six and twelve carbon atoms (Tables 8.3 and 8.4).

Although ethanal, propanal and 2-propanone were the dominant carbonyls detected in a Swedish fermented sausage (Halvarson, 1973), it was concluded that they probably had little affect on aroma since their concentrations changed little during ripening of either the air-dried or smoked product he investigated. On the other hand, increases in concentration of octanal, 2-hexenal and 2-heptenal, which were particularly evident in the air-dried sausage, were thought to be of much more relevance. They have flavour thresholds of between 5 and 60 ppb and flavours variously described as 'green', 'metallic', 'fruity' and 'fatty' (Forss, 1972). The negative influence of smoking on the formation of these compounds was interpreted to be consistent with formation of the carbonyls by autoxidative processes in view of the known anti-oxidative properties of smoke components such as phenols. Of the other carbonyls mentioned as being of importance by Halvarson (1973), viz C_6 and C_7 dienals, both decreased in concentration during ripening of the air-dried product, while the C_6 compound increased and the C_7 decreased in concentration in the smoked product.

8.6 Headspace volatile compounds

Many carbonyl compounds were detected by gas chromatography–mass spectrometry (GC–MS) of the headspace volatiles of air-dried, mould-fermented, Italian-style salamis (Tables 8.3 and 8.4, column 5). The absence of dienals (cf. above) was interpreted by the authors (Berger et al., 1990) as evidence against autoxidative, and for microbial oxidation of fatty acids. They suggested peroxidation of fatty acids by hydrogen peroxide-producing lactobacilli, with 'control' against excessive oxidation by catalase-producing Micrococcaceae strains. The latter, together with moulds, are most numerous at or near to the sausage surface and are said to minimise oxygen penetration through their own oxygen requiring metabolism (Lücke, 1986). The presence of anti-oxidative compounds from the garlic and pepper constituents of the sausage were cited as another argument against autoxidation (Berger et al., 1990). In view of the existence of dienal-degrading organisms (Alford, 1971) and the presence of so many other characteristic carbonyls, equally convincing arguments for autoxidation can be put forward.

Whatever their mechanisms of formation, Berger et al. (1990) concluded that fat-degradation products were one of two important elements in the odour profile of their salami samples, the other being volatiles derived from spices (see below and Table 8.5).

In a second GC–MS study of headspace volatiles, this time associated with spice-free sausage mixes inoculated with various combinations of starter cultures (Monteil, 1991), presumed lipid oxidation products made up 60% of the volatiles recovered. Included, in addition to the 2-alkanones (C_3–C_8) and aldehydes (C_5–C_9) recorded by many previous investigators, were straight-chain, saturated and unsaturated hydrocarbons (C_5–C_{10}), three furans (Tables 8.3 to 8.5) and a lactone. Statistical analysis of headspace and sensory data, showed that starter cultures comprising *Staphylococcus carnosus* with any one of *Lactobacillus sake*, *Pediococcus acidilactici* or *P. pentosaceus* produced the highest amounts of the C_5–C_7 2-alkanones, of 3-methylbutanal and 3-methylbutanol and the most marked 'cured' odours.

The sources of the two branched-chain compounds and the related acids (Table 8.1) was assumed to be microbial metabolism of the corresponding amino acids. Similarly, cystine-cysteine and phenylalanine were proposed as the sources of dimethyldisulphide and benzene acetaldehyde, respectively. The presumed end-products of amino acid metabolism accounted for some 6% of the total volatiles.

Table 8.5 Compilation from published data of 'non-carbonyl' compounds present in headspace volatiles of fermented sausages

Compound	1	2	3	Compound	1	2	3
Aliphatic alcohols				*Alphatic esters*			
Ethanol		×		Ethyl acetate		×	×
Propan-1-ol		×		Ethyl propanoate			×
Butan-1-ol			×	Ethyl lactate			×
Penetan-1-ol	×		×	Ethyl butanoate			×
Hexan-1-ol		×	×	Ethyl pentanoate			×
Heptan-1-ol			×	Ethyl hexanoate			×
Octan-1-ol			×	Ethyl heptanoate			×
Butan-2-ol			×	Ethyl ocatanoate			×
Pentan-2-ol				Ethyl hexenoate			×
Heptan-2-ol			×	Ethyl pentenoate			×
1-Penten-3-ol		×		Ethyl 2-methylpropanoate			×
3-Penten-2-ol		×		Ethyl 2-methylbutanoate		×	
Hexen-1-ol			×	Ethyl 3-methylbutanoate			×
Hexen-2-ol	×			Ethyl methylpentanoate			×
1-Octen-3-ol	×	×	×	Propyl acetate		×	
2-Methylpropan-1-ol			×	Hexyl acetate			×
2-Methylbutan-1-ol			×	2-Methylpropyl acetate			×
3-Methylbutan-1-ol		×	×	2-Methylpropyl propanoate			×
2-Ethylhexan-1-ol			×	3-Methylbutyl acetate			×
3-Methylbutan-2-ol	×						
3-Methyl-1-hexen-3-ol	×			*Nitrogen-containing compounds*			
				2,6-Dimethylpyrazine		×	×
Hydrocarbons				2,3,5-Trimethylpyrazine			×
Pentane		×		Organic nitrates			×
Hexane		×					
Heptane		×		*Sulphur-containing compounds*			
Octane		×		3,3′-Thiobis-1-propene			×
Nonane		×	×	Methyl 2-propenyl disulphide			×
1-Heptene		×		Di-2-propenyl disulphide			×
1-Octene		×		Dimethyldisulphide		×	
1-Nonene			×	Thietane		×	
2-Octene		×		Unknown		×	
4-Octene		×					
4-Methyl-1-pentene		×					
2,2,3-Trimethylnonane		×					
				Furans			
Terpenes†	×	×	×	Tetrahydrofuran		×	
				2-Ethyl furan		×	
Aromatic compounds	×	×	×	2-Pentyl furan		×	

* 1. Berger *et al.*, 1990. 2. Montcil, 1991. 3. Edwards *et al.*, 1991.
† >20 compounds in study 1; 1 in study 2; 15 in study 3.

In the same study, *Staph. warneri* or *Staph. saprophyticus* combined with *L. sake* or *P. acidilactici* produced the highest amounts of acetoin (3-hydroxybutan-2-one), diacetyl (butan-2,3-dione), the structurally related 1,3- and 2,3-butandiols and acetic acid (Tables 8.1 and 8.5). The dominating odour was now 'buttery', in keeping with published sensory data for acetoin and diacetyl (Forss, 1972). The source of these compounds, which together with ethanol and acetaldehyde (Table 8.5) represented some 27% of the total volatiles, was assumed to be microbial carbohydrate fermentation.

Alkanes, alcohols (except ethanol) and straight-chain aldehydes (Table 8.5) were found in greatest amounts whenever *Staph. warneri* was used. Alkanes have relatively high flavour threshold concentrations in the ppm range, while primary and secondary saturated and unsaturated alcohols (C_7–C_{10}) have thresholds less than 1 ppm. Particularly potent is 1-octen-3-ol with a detection threshold between 1 and 10 ppb (Forss, 1972); it was detected not only in the French study but in the other two studies for which results are given in Table 8.5 as well.

In addition to aldehydes, ketones and a relatively large number of alcohols, the samples of Spanish *salchichon*, *fuet*, and *chorizo* sausages examined in a third GC–MS study (Edwards *et al.*, 1991), contained a series of esters. These were mainly ethyl esters of saturated, unsaturated and branched-chain acids with up to eight carbon atoms (Table 8.5). Although concentrations were not calculated, the ion current intensities from MS indicated the esters were major components of the headspaces.

Using a technique to produce sausage from minimally microbially contaminated ingredients (Ordonez *et al.*, 1989), microbial involvement in ester formation appeared to be essential, though no specific bacterial source was found. On the other hand, Monteil (1991) showed that a starter culture comprising *L. sake* and *Staph. saprophyticus* produced the highest levels of the two esters detected in her study (Table 8.5). Two of the four esters detected by Wittkowski *et al.* (1990) in *landjäger* salami, i.e, ethyl acetate and methyl butanoate, were also present, along with eight others, in the sample of commercial liquid smoke they analysed. Two others, the ethyl esters of butanoic and 2-methyl propanoic acids were not. In this particular study pepper volatiles dominated a relatively simple mixture of headspace volatiles comprising ethanol, 2-propanone, 2-hexanone, butan-2,3-dione, 3-methylbutanal, furan, 2-methyl furan, hexane, benzene and the esters mentioned.

All esters have highly characteristic fruity odours with flavour thresholds typically in the 10–50 ppb range (Forss, 1972). Many years ago, Keller (1955) reported the possible importance to flavour of fermented sausages of compounds produced by Gram-negative bacteria isolated from sausages ripened at 12°C which had a *Fruchtester-Aroma*. Whether these compounds were esters or other compounds with similar odours, e.g. some carbonyls, was not determined. A range of esters including the ethyl derivatives of 2-methylbutanoic, hexanoic and octanoic acids (cf. Table 8.5) and six other undefined alkyl esters of longer-chain acids have been detected amongst the headspace volatiles of Spanish Iberian hams (Garcia *et al.*, 1991). Amongst the volatiles of French dry-cured hams, just a single ester, methyl decanoate, was detected together with a range of free fatty acids including the C_5 C_6 and C_8 esterified acids in Iberian ham (Berdague *et al.*, 1991). In both studies, the esters were components of complex mixtures of branched and unbranched hydrocarbons, aldehydes and alcohols, and mono- and diketones, i.e. showing overall similarities in composition to the respective fermented sausage volatiles. There were many qualitative and quantitative differences in individual components, however.

To all of the processes and end-products already discussed must be added the contributions from spices and/or smoke whose use is of course very much product specific. Terpenes believed to be derived from pepper, were prominent amongst the volatiles detected in three of the four studies of headspace volatiles mentioned above (Berger *et al.* (1990), Wittkowski *et al.* (1990) and Edwards *et al.* (1991). In the other (Monteil, 1991), spices were intentionally omitted from the ingredients so the single terpene detected, limonene, was presumably a contaminant.

Berger *et al.* (1990) concluded from the effects of certain further chemical treatments on their samples that microbes were involved in the degradation of pepper terpenes to a range of shorter-chain hydrocarbons, which were present in their samples. The three sulphur-containing compounds listed for the Spanish sausages in Table 8.5 were only recovered from *chorizo* samples and were probably derived from the garlic characteristically added to this particular product. Such compounds and/or terpenes, when present in the sort of amounts typically used in fermented sausages will certainly have an important and perhaps dominating influence on the sensory properties of the product.

The importance of nitrite to flavour of fermented sausages has been shown by the sensory experiments of Noel *et al.* (1990), in which products fermented without nitrite had inferior sensory qualities to the equivalent product with nitrite. Flavour compounds presumably arise from reactions between the added, or microbially produced, nitrite and other sausage components. The organic nitrates found in some Spanish products (Table 8.5) were probably of such origin as shown for the same type of compounds detected in bacon (Mottram *et al.*, 1984).

Finally, the importance of sodium chloride must not be forgotten amongst the findings of so many other 'exotic' organic chemicals

As indicated at the beginning of the chapter, it has only been possible to catalogue the processes and end-products of potential relevance to flavour development in fermented sausages and speculate on their importance. Much more combined chemical and sensory studies such as that of Monteil (1991) using minimally contaminated ingredients, different starter cultures and controlled conditions of ripening are now needed if further progress is to be made.

References

Alford, J.A. (1971) Relationship of microbial activity to changes in lipids of foods. *J. Appl. Bact.*, **34**, 133–146.

Astiasaran, I., Villanueva, R. and Bello, J. (1990) Analysis of proteolysis and protein insolubility during the manufacture of some varieties of dry sausage. *Meat Sci.*, **28**, 111–117.

Berdague, J.-L., Denoyer, C., Le Quere, J.-L. and Semon, E. (1991) Volatile components of dry-cured ham. *J. Agric. Food Chem.*, **39**, 1257–1261.

Berger, R.G., Macku, C., German, J.B. and Shibamoto, T. (1990) Isolation and identification of dry salami volatiles. *J. Food Sci.*, **55**, 1239–1242.

Cantoni, C., Bianchi, M.A. and Beretta, G. (1974) Variazioni di aminoacidi liberi istamina e tiramina in insaccati stagionati (salami). *Industrie Alimentari*, **13**, 75–78.

Cantoni, C., Molnar, M.R., Renon, P. and Giolitti, G. (1966). Investigations on the lipids of dry sausages. In *Proc. 12th Meeting Eur. Meat Workers, Sandefjord*, paper E-4.

Cerise, L., Bracco, U., Horman, I., Sozzi, T. and Wührmann, J.J. (1973) Veränderungen des Lipidanteils während des Reifungsprozesses von Salami aus reinem Schweinefleisch. *Fleischwirtschaft*, **53**, 223–225.

Dainty, R.H. and Hibbard, C.M. (1983) Precursors of the major end-products of aerobic metabolism of *Brochothrix thermosphacta*. *J. Appl. Bact.*, **55**, 127–133.

De Ketelaere, A., Demeyer, D., Vandekerckhove, P. and Vervaeke, I. (1974) Stoichiometry of carbohydrate fermentation during dry sausage ripening. *J. Food Sci.*, **39**, 297–300.

DeMasi, T.W., Wardlaw, F.B., Dick, R.L. and Acton, J.C. (1990) Nonprotein nitrogen (NPN) and free amino acid contents of dry, fermented and nonfermented sausages. *Meat Sci.*, **27**, 1–12.

Demeyer, D., Hoozee, J. and Mesdom, H. (1974) Specificity of lipolysis during dry sausage ripening. *J. Food Sci.*, **39**, 293–296.

Demeyer, D., Vandekerckhove, P. and Moermans, R. (1979) Compounds determining pH in dry sausages. *Meat Sci.*, **3**, 161–167.

Demeyer, D., Verplaetse, A. and Gistelinck, M. (1986) Fermentation of meat: an integrated process. In *Proc. 32nd Eur. Meeting Meat Res. Workers, Ghent*, pp. 241–247.

Dierick, N., Vandekerckhove, P. and Demeyer, D. (1974) Changes in nonprotein nitrogen compounds during dry sausage ripening. *J. Food Sci.*, **39**, 301–304.

Dobbertin, S., Siems, H. and Sinell, H.-J. (1975) Beiträge zur Bakteriologie der frischen Mettwurst. II Mitteilung: Die Abhängigkeit lipolytischer Aktivität von der Keimdynamik in frischen Mettwürsten. *Fleischwirtschaft*, **55**, 237–242.

Dominguez Fernández, M.C. and Zumalacárregui Rodriguez, J.M. (1991) Lipolytic and oxidative changes in *chorizo* during ripening. *Meat Sci.*, **29**, 99–107.

Dwivedi, B.K. (1975). Meat flavour. *CRC Crit. Rev. Food Technol.*, **5**, 487–535.

Edwards, R.A., Dainty, R.H., Hibbard, C.M. and Ramantanis, S.V. (1987) Amines in fresh beef of normal pH and the role of bacteria in changes in concentration observed during storage in vacuum packs at chill temperatures. *J. Appl. Bacteriol.*, **63**, 427–434.

Edwards, R.A., Dainty, R.H. and Ordonez, J.A. (1991) Volatile compounds of microbial origin in dry, fermented Spanish sausage. Poster, 2nd International Symposium on the Interface between Analytical Chemistry and Microbiology–Chromatography and Mass Spectrometry in Microbiology. Lund, Sweden.

Forss, D.A. (1972). Odour and flavour compounds from lipids. In *Progress in the Chemistry of Fats and other Lipids*. Ed. R.T. Holman, pp. 177–258. Pergamon Press, Oxford.

Garcia de Fernando, G.D. and Fox, P.F. (1991) Study of proteolysis during the processing of a dry fermented pork sausage. *Meat Sci.*, **30**, 367–383.

Garcia, C., Berdagué, J.J., Antequera, T., López-Bote, C., Córdoba, J.J. and Ventanas, J. (1991) Volatile components of dry cured Iberian ham. *Food Chem.* **41**, 23–32.

Grazia, L., Romano, P., Bagni, A., Roggiani, D. and Guglielmi, G. (1986) The role of moulds in the ripening process of salami. *Food Microbiol.* **3**, 19–25.

Halvarson, H. (1973) Formation of lactic acid, volatile fatty acids and neutral, volatile monocarbonyl compounds in Swedish fermented sausage. *J. Food Sci.* **38**, 310–312.

Kandler, O. (1983) Carbohydrate metabolism in lactic acid bacteria. *Antonie van Leeuwenhoek*, **49**, 209–224.

Keller, H. (1955) Versuche zür Bakteriellen Aromatisiering der Rohwurst. *Ann. Inst. Pasteur (Lille)* 7, 235–239.

Langner, H.J. (1969) Zur Bildung von freie Aminosauren, flüchtigen Fettsauren und flüchtigen Carbonylen in reiender Rohwurst. *Fleischwirtschaft*, **49**, 1475–1479.

Langner, H.J. (1972) Aromastoff in der Rohwurst. *Fleischwirtschaft*, **52**, 1299–1306.

Lois, A.L., Gutierrez, L.M., Zumalacarregui, J.M. and Lopez, A. (1987) Changes in several constituents during the ripening of *Chorizo* — A Spanish dry sausage. *Meat Sci.* **19**, 169–177.

Lücke, F.-C. (1986) Microbiological processes in the manufacture of dry sausage and raw ham. *Fleischwirtschaft*, **66**, 1505–1509.

Lücke, F.-C. and Hechelmann, H. (1986) Starterkulturen fur Rohwurst und Rohschinken. *Fleischwirtschaft*, **66**, 154–166.

Monteil, P. (1991) Incidence des ferments d'ensemencement sur la formation des composes d'aromes dans le saucisson sec. Thesis, Universite Blaise Pascal, Clermont-Ferrand, France.

Mottram, D.S. (1991) Meat. In *Volatile Compounds in Foods and Beverages*, Ed. H. Maarse, pp. 107–177. Marcel Dekker, New York.

Mottram, D.S., Croft, S.E. and Patterson, R.L.S. (1984) Volatile components of cured and uncured pork: the role of nitrite and the formation of nitrogen compounds. *J. Sci. Food Agric.*, **35**, 233–239.

Nagy, A., Mihalyi, V. and Incze, K. (1989) Ripening and storage of Hungarian salami. *Fleischwirtschaft*, **69**, 587–588.

Noel, P., Briand, E. and Dumont, J.P. (1990) Role of nitrite in flavour development in uncooked cured meat products: Sensory assessment. *Meat Sci.*, **28**, 1–8.

Nurmi, E. & Niinivaara, F.P. (1964) Lipolytic changes of fats in dry sausages. In *Proc. 10th Eur. Meeting Meat Res. Workers, Roskilde*, Denmark, paper G-8.

Ordonez, R.A., Asensio, M.A., Garcia, M.L., Dolores Selgas, M. and Sanz, B. (1989) A resonably aseptic method of monitoring the phenomena occurring during the ripening of dry fermented sausages. *Fleischwirtschaft*, **69**, 1023–1025.

Pezacki, W, and Pezacka, E. (1983) Einflüss der Salzung von Rohwurstbrat auf die Proteolyse. *Fleischwirtschaft*, **63**, 625–630.

Reuter, G., Langner, H.J. & Sinell, H.-J. (1968) Entwicklung der Mikroflora in schnellreifender deutscher Rohwurst and analoge quantitative Aminosaureanalyse bei einer Salami. *Fleischwirtschaft*, **48**, 170–176.

Roncales, P., Aguilera, M., Beltran, J.A., Jaime, I. and Peiro, J.M. (1989) Effect of the use of natural or artificial casings on the ripening and sensory quality of dry sausage. In *Proc. 35th Int. Congr. Meat Sci. Techn. Copenhagen*, pp. 825–832.

Sajber, C., Karakas, R. and Mitic, P. (1971) Influences of some starter cultures upon the changes in properties of *stajer* sausages during the fermentation. In *Proc. 17th Eur. Meeting Meat Res. Workers, Bristol*, pp. 744–760.

Tiecco, G., Tantillo, G., Marcotrigiano, G. and De Natale, G. (1985) Determinazione dell'istimina ed altre amine biogene con la cromatografia liquida ad alta risoluzione negli insaccati. *Industrie Alimentari*, **24**, 122–126.

Vandekerckhove, P. (1977) Amines in dry fermented sausage, *J. Food Sci.*, **42**, 283–285.

Verplaetse, A., De Bosschere, M. and Demeyer, D. (1989) Proteolysis during dry sausage ripening. In *Proc. 35th Int. Congr. Meat Sci. Technol. Copenhagen*, pp. 815–818.

Wahlroos, O. and Niinivaara, F.P. (1969) Chemical changes in lipids and sulphur-containing substances during ripening of raw sausage. In *Proc. 15th Eur. Meeting Meat Res. Workers, Helsinki*, pp. 240–251.

Wittkowski, R., Baltes, W. and Jennings, W.G. (1990) Analysis of liquid smoke and smoked meat volatiles by headspace gas chromatography. *Food Chem.*, **37**, 135–144.

9 Fungal toxins in raw and fermented meats

J. PESTKA

9.1 Introduction

Fungi impact human health directly or indirectly via mycoses, allergies, and mycotoxicoses. A mycosis is as an invasion of living tissue by a fungus, whereas allergies are manifestations of a hyperactive immune response induced by fungal cell components. In contrast, mycotoxicoses are toxic effects in animals that result from ingestion of fungal secondary metabolites known as mycotoxins. While mycoses and allergies to fungi have not been a major concern in meat consumption, mycotoxins theoretically can persist in meats as tissue residues carried over from contaminated feeds or occur as a result of toxinogenesis during meat fermentations that characteristically include mold growth. Numerous mycotoxins have been identified during the last thirty years. Some of these compounds such as aflatoxin B_1 are potent carcinogens, while others impair normal digestive, reproductive, neurologic or immunologic function (Pestka and Casale, 1990; Pestka and Bondy, 1989). Significant questions exist over the extent to which meat-borne mycotoxins pose a health threat and these impact on both approaches to animal husbandry and food technology. This chapter describes those mycotoxins of primary concern to human health, reviews their capacity to be transmitted to edible animal tissues, and discusses their potential occurence in molded meats.

9.2 Mycotoxins: occurrence, toxicity and tissue carryover

Mycotoxins are a heterogeneous group of non-polar low-molecular-weight compounds produced by fungi as part of secondary metabolism. Toxinogenic fungi are ubiquitous and can grow in various agricultural commodities in the field, or, appear postharvest during storage and processing. It is important to note that not all fungi are toxinogenic and that production of mycotoxins by toxinogenic fungi is not always concurrent with growth but, rather, is dependent on the temperature, relative humidity, and water content of the substrate. Thus, weather and storage conditions often dictate whether mycotoxins will be present in a commodity. Many mycotoxins have been discovered in association with livestock and poultry toxicoses. Often fungi are isolated from feed associated with an animal toxicoses of unknown etiology and the organisms then shown to produce toxic components in culture. These compounds have been systematically extracted, purified, and chemically characterized using cellular and animal biossays. Based on these investigations and human epidemiological evidence, potential harmful effects of mycotoxins are sometimes extrapolated to humans. Mycotoxins that have possible significance to meat products include the aflatoxins, ochratoxins, cyclopiazonic acid, trichothecenes, zearalenone, and the fumonisins. Table 9.1 summarizes sources and toxic manifestations of these and other mycotoxins produced by *Aspergillus*, *Penicillium* and *Fusarium*.

9.2.1 Aspergillus *and* Penicillium *toxins*

Aflatoxins. The aflatoxins are a family of difuranocoumarins, produced by toxinogenic strains of *Aspergillus flavus* and *A. parasiticus*. Because aflatoxins are hepatotoxic and hepatocarcinogenic, these compounds have received notoriety and their potential presence in animal tissue has been of great concern. Aflatoxins were first identified during the early 1960s in association with an epidemic known as turkey X disease in which over 100 000 turkey poults died after ingesting toxic peanut meal (Austwick, 1978). It was concurrently discovered that aflatoxin contamination of cottonseed meal was the cause of trout hepatoma epidemics in hatcheries in the western USA (Sinnhuber *et al.*, 1965).

Under appropriate humidity and temperatures, toxinogenic aspergilli grow and produce aflatoxins on almost any organic substrate. Agricultural commodities that can be contaminated by the aflatoxins include corn, peanuts, cottonseed, cassava, copra, and many types of treenuts (Davis and Deiher, 1987). Figure 9.1 shows the structures of four naturally occurring aflatoxins, aflatoxin B_1 (AFB$_1$), aflatoxin B_2 (AFB$_2$), aflatoxin G_1 (AFG$_1$) and aflatoxin G_2 (AFG$_2$), of which AFB$_1$ is encountered most often in nature.

Aflatoxins are toxic, carcinogenic, mutagenic, and teratogenic (Busby and Wogan, 1979). The rank order for toxicity, carcinogenicity, and mutagenicity among the four naturally occurring aflatoxins is AFB$_1$ > AFG$_1$ >AFB$_2$ > AFG$_2$, indicating that the unsaturated terminal furan of AFB$_1$ is critical for determining the degree of biological activity of this mycotoxin family. Upon acute AFB$_1$ exposure, toxicity occurs in 3–6 hours and is evidenced by necrosis of the hepatocytes, disruption of the normal clotting mechanisms, and capillary fragility resulting in hemorrhaging and death (Cysewski *et al.*, 1968, Carlton and Szczech 1978). An outbreak of acute aflatoxin toxicity occurred in India during 1974 when 106 people died from consumption of moldy, aflatoxin-contaminated corn (Van Rensburg, 1977). Aflatoxins also cause marked decreases in productivity and in the

Table 9.1 Major foodborne mycotoxins and their effects*

Mycotoxin	Species affected	LD$_{50}$ range (mg/kg)	Effects
Aflatoxins	Rodents, poultry, swine, cattle, humans	0.4–18 (AFB$_1$), 0.8 (AFG$_1$), 170 (AFG$_2$), 17 (AFM$_1$)	Hemorrhage Hepatotoxicity Hepatocarcinogenicity
Citrinin	Rodents, swine, dog	19–67	Nephrotoxicity Porcine nephropathy
Cyclopiazonic acid	Rodents, poultry, swine	2.3–70	Muscle necrosis Intestinal hemorrhage Edema Oral lesions
Fumonisin	Rodents, swine, horses	–†	Leukoencephalomacia Pulmonary edema Hepatocarcinogenicity Tumor promotion
Ochratoxin A	Rodent, poultry, swine, dog, human	5.5–22	Nephrotoxicity Porcine nephropathy Enteritis Renal and urinary-tract cancer Immunotoxic
Patulin	Rodents, poultry, cattle, cat	8–50	Edema Hemorrhage Convulsions
Penicillic acid	Rodents, poultry	100–300	Hepatotoxicity Nephrotoxicity Edema Carcinogenesis
Sterigmatocystin	Rodents	32–166	Hepatotoxic Hepatocarcinogen
Trichothecenes	Rodents, poultry, cattle, swine, dog, cat, human	1.8–5.2 (T-2), 70–77 (deoxynivalenol), 0.8–23 (diacetoxyscirpenol)	Digestive disorders Hemorrhage Edema Oral Lesions Immunotoxicity
Zearalenone	Rodents, poultry, swine, cattle	–†	Estrogenic effects Testicular atrophy Abortion Infertility

* Adapted from CAST (1989).
† Low level of acute toxicity.

immune response of domestic animals at chronic levels (CAST, 1989). Reduced growth rate in livestock and poultry is the most common effect of aflatoxin in feed.

Rat, monkey, marmoset, ferret, trout, and salmon are among the species where experimental AFB-induced hepatocarcinogenesis has been demonstrated. Rainbow trout are one of the most sensitive of the species;

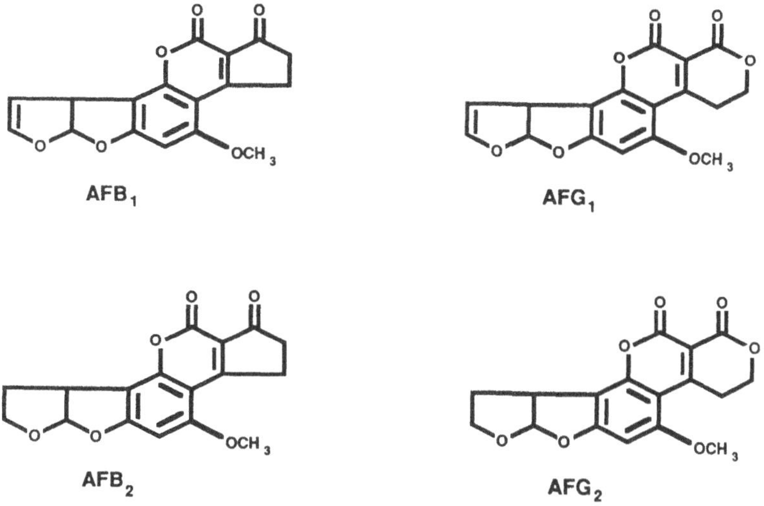

Figure 9.1 Naturally occurring aflatoxins.

as low as 4 ppb of AFB_1 in the diet for 12 months is adequate to cause liver tumors in test animals (Sinnhuber *et al.*, 1968). Epidemiological studies have suggested that a positive correlation exists between current exposure to aflatoxins in foods and the incidence of primary human liver cancer in areas of Africa, India, South-East Asia, and China (Shank, 1978). However, these populations are also at high risk of exposure to hepatitis B virus which also contributes to human liver cancer.

Aflatoxins, like other non-polar xenobiotics, are metabolized primarily in the liver by cytochrome P450 enzymes, further detoxified by conjugation with glucuronic acid or sulfate and eliminated via urine and faeces. Figure 9.2 shows the structures of some major AFB_1 metabolites. AFM_1 which results from hydroxylation of AFB_1 at the 4-position, has about 3% of the mutagenic potency of AFB_1 in the *Salmonella* mutagenesis assay (Wong and Hsieh, 1976) and is also less toxic and carcinogenic. Animals which have ingested AFB_1 in their diet at a level of 300 ppb will produce milk containing 1 ppb AFM_1 24 hours later. Detectable AFM_1 disappears 4–5 days after withdrawal from AFB_1-contaminated feed. Government action levels as low as 0.05 ppb for AFM_1 have been established for milk and dairy products because of the potential danger to growing children who ingest large quantities of milk (Van Egmond, 1989). Aflatoxicol (AFL) results from the reversible reduction of the cyclopentenone carbonyl of AFB_1 by cytoplasmic enzymes. Formation of AFL is not considered to be a true detoxification mechanism since this metabolite has about 23% the mutagenic potency of AFB_1 (Wong and Hsieh, 1976) and is nearly as carcinogenic as AFB_1 in trout and rats (Nixon *et al.*, 1981; Schoenhard *et*

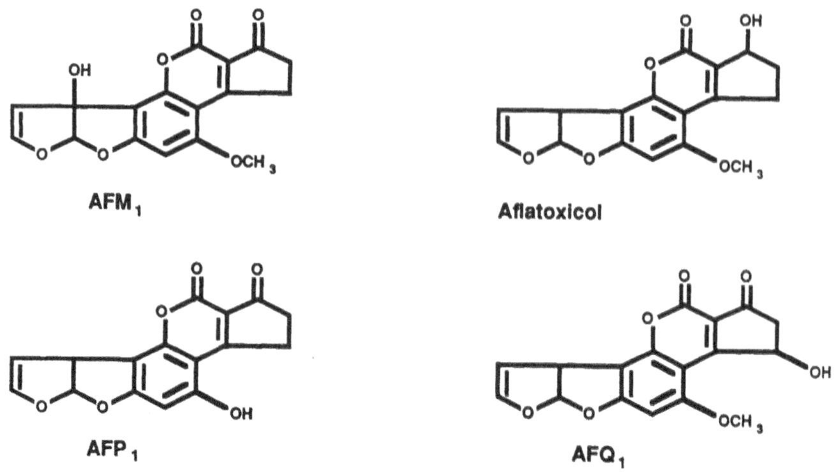

Figure 9.2 Selected aflatoxin B metabolites

al., 1981). Two other major AFB_1 metabolites, AFQ_1, resulting from ring hydroxylation of the carbon B to the cyclopentenenone carbonyl, and AFP_1, resulting from *o*-demethylation, have 1 and 0.1%, respectively, the mutagenic potency of the parent aflatoxin.

The carcinogenic and mutagenic potency of AFB_1 results from the formation of a reactive epoxide at the 2,3-position of the terminal furan and its subsequent covalent binding to nucleic acid (Essigmann *et al.*, 1982). Some nucleic acid adducts can last for weeks after initial formation. An important means of detoxifying the active AFB_1 epoxide is via formation of a glutathione conjugate (Moss *et al.*, 1983). Another major metabolite, 2, 3-dihydro-2, 3-dihydroxyaflatoxin B_1-diol appears after spontaneous or enzymatic reaction of the 2, 3-epoxide with water (Lin *et al.*, 1977; Neal and Colley, 1979) and after the degradation of AFB_1-modified DNA (Wang and Cerutti, 1980). A functional role for AFB_1-diol in aflatoxicosis has been postulated on the basis that it forms adducts with amino groups of cellular proteins (Neal *et al.*, 1981).

AFB_1 and its metabolites may to a certain extent accumulate in animal tissue and be carried into meat products (Rodricks and Stoloff, 1977; Stoloff, 1979; Pestka, 1986). Critical factors affecting aflatoxin residue accumulation in tissue include species of animal, level of exposure, health status of animal, and the time after cessation of aflatoxin exposure that the tissues are analysed. Most ^{14}C-labeled AFB_1 injected intraperitoneally into rats is eliminated within 24 hours in the urine and faeces with about 8% remaining in the liver (Wogan *et al.*, 1969). Luthy *et al.* (1980) performed a comprehensive study in which excretion of ^{14}C-labeled AFB_1 and its metabolites was followed for 9 days in pigs. Approximately 71% of the label was released in the faeces and urine, of which only 11% was extractable in methylene chloride. Tissue analyses revealed that the highest ^{14}C activities were in the liver followed by kidney and lung, most of which was neither extractable or dialyzable. General conclusions from this and earlier tracer studies were: (i) a major portion of administered AFB_1 is rapidly metabolized to water-soluble conjugates (probably glucuronide, sulfate or glutathione metabolites), which are excreted in urine and faeces; (ii) most of the residues remaining in tissues and plasma are in a bound form; and (iii) only small fractions of residues in faeces, urine, and tissue, are actually AFB_1 or biologically active metabolites, such as aflatoxicol, AFM_1 or AFQ_1.

Several investigators have identified and quantitated specific aflatoxin residues in tissues from swine fed contaminated diets (Jacobson *et al.*, 1978; Neff and Edds, 1981; Furtado *et al.*, 1982; Trucksess *et al.*, 1982). Although these data vary slightly because of different detection methods, the general conclusions remain the same. Pigs fed diets containing between 0.1 and 1.3 ppm AFB_1 for various periods of time have detectable amounts of AFB_1 and AFM_1 in their tissues at the immediate conclusion of the experiments. Total residues appear to be dependent on dose levels but independent of dose duration. Primary sites of accumulation are liver and kidney, with maximal reported levels approaching 1–2 ppb of both AFB_1 and AFM_1. Accumulation in muscle is much less than in liver or kidney. These results are not surprising since the latter two organs are primary sites of xenobiotic metabolism and elimination. One study that analysed AFL in swine found that the metabolite accounted for only 1% AFB_1 + AFM_1 accumulation in liver but that it was not detectable in kidney or muscle (Trucksess *et al.*, 1982). An important conclusion from these studies was that low-level aflatoxin residues can be greatly reduced by withdrawing pigs from contaminated feed and placing them on clean feed for a single day. Furthermore, residues are completely eliminated by a four-day withdrawal period (Furtado *et al.*, 1982; Gregory *et al.*, 1983).

Assessment of aflatoxin residues in edible tissues of cows indicates, as with swine, that more AFB_1, and AFM_1 accumulate in kidney and liver than in muscle (Richard *et al.*, 1983; Stubblefield *et al.*, 1983; Trucksess *et al.*, 1983a). Dietary and oral administration studies reveal that the kidney accumulates the most residues and that these are primarily AFM_1 with lesser amounts of AFB_1 and AFL. The AFB_1 and AFM_1 residues remain in the liver and kidney seven days after oral dosing but are completely eliminated from cattle tissue within 18 days after withdrawal from the contaminated diet. It appears that the cow might require a longer withdrawal period than the pig for complete elimination of aflatoxin residues from tissues, although further studies are needed to clarify this.

The risk from ingestion of water-soluble and covalently bound macromolecular aflatoxin forms might also be questioned. Although also present at very low levels in animal tissue, free aflatoxins may be cleaved from

Table 9.2 Carryover of AFB_1 in feed to AFB_1 and AFM_1 in animal tissue

| Species | Feed:tissue ratio* (AFB_1:AFB_1 + AFM_1) | | | Reference |
	Liver	Kidney	Muscle	
Pig	1 316	1 103	–†	Neff and Edds (1981)
Pig	489	467	1 343	Furtado et al. (1982)
Pig	251	444	938	Miller et al. (1982)
Pig	323	250	200 000	Trucksess et al. (1982)
Steer	268	79	3 974	Richard et al. (1983)
Turkey	926	–	16 667	Gregory et al. (1983)
Chicken	19 512	30 769	–	Trucksess et al. (1983b)
Chicken	8 943	13 713	18 700	Chen et al. (1984)

* Based on AFB_1 in feed and tissue analyses within 24 hours after completion of feeding regimen.
† Not determined.

glucuronide, sulfates, nucleic acids, and proteins during digestion and subsequently absorbed from the intestine. Jaggi *et al.* (1980) tested this hypothesis by preparing *in vivo* water-soluble and macromolecular-bound forms of radiolabeled AFB_1 and feeding these to rats. Based on radioactivity covalently bound to liver DNA (a measure of potential carcinogenicity), the macromolecule-bound AFB_1 derivatives were at least 4000 times less active than AFB_1 with respect to covalent binding to rat liver DNA, and the water-soluble conjugates were at least 100 times less potent than free AFB_1. It was concluded by the authors that the risk of cancer to humans who consume liver or meat containing such aflatoxin residues is negligible when compared to the risk from intake of aflatoxins from nuts, milk, and other sources.

Table 9.2 summarizes the estimated feed:tissue ratios of AFB_1 to AFB_1 and AFM_1 in pig, steer, turkey and chicken. In general, feed:tissue ratios are much lower for kidney and liver than for muscle. Chickens apparently have the least ability to carry over AFB_1 into tissue. Bovine kidney has the lowest feed:tissue ratio (79) of all the organs examined among the species. Because of improved analytical techniques, these values are generally lower than those discussed in an earlier review by Rodricks and Stoloff (1977).

Cooking has little effect on aflatoxins in meat (Furtado *et al.*, 1981). Although it is often not economically feasible to preclude the use of feeds contaminated with low-to-moderate amounts of aflatoxins, several alternatives are available (CAST, 1989). For example, withdrawal of aflatoxin-contaminated feed could be an effective means of eliminating aflatoxin residues from the tissues of food-producing animals. However, it would not be practical to routinely implement such a control programme among livestock and poultry producers. Another approach for decreasing aflatoxin residues would be to detoxify contaminated feed by a chemical or physical method. Ammoniation of feed has been shown to be very effective

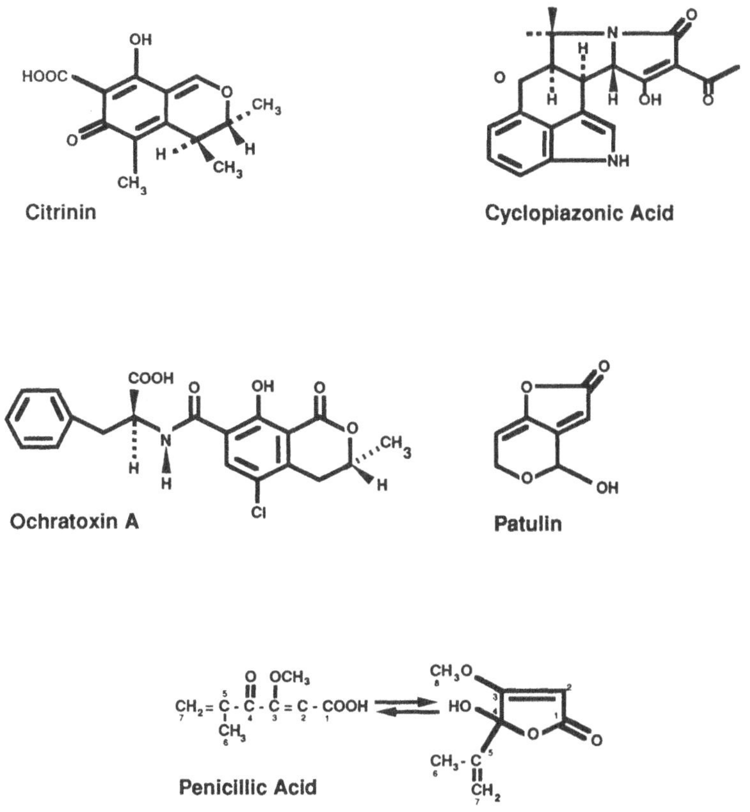

Figure 9.3 Selected *Aspergillus* and *Penicillium* toxins.

in decreasing total aflatoxins in foods (Park and Rua, 1990). While the process is approved in some countries, there are questions regarding the identity or toxicity of substances formed by the ammoniation of aflatoxins. An alternative possibility is to modify the bioavailability of aflatoxin by incorporating a chemosorbent in feed such as hydrated sodium calcium aluminosilicate (Phillips *et al.*, 1988). In considering these and other possibilities, it will ultimately be necessary to balance the minimal health impact of very low levels of aflatoxins in tissues against the economic benefits of using contaminated feed.

Ochratoxins. The ochratoxins are a group of seven related isocoumarin derivatives linked to phenylalanine that are produced by species of *Aspergillus* and *Penicillium* (Busby and Wogan, 1979). Ochratoxin A (OA), the most commonly encountered of the group (Figure 9.3), is nephrotoxic to monogastric animals and has also been shown to be carcinogenic and immunotoxic (Kuiper-Goodman and Scott, 1989; Pestka

Table 9.3 Carryover of ochratoxin A into edible animal tissue

Species	Dietary (ppb)	Dose interval	Withdrawal time (days)	Tissue (ppb) Liver	Kidney	Muscle	Reference
Pig	1000	3 months	0	11.4	26.8	9.5	Krogh et al. (1979)
Calf	430	87 days	0	n.d.*	trace–5	–†	Patterson et al. (1981)
Chicken	500	14 days	0	13	124	3.3	Frye and Chu
	5000	14 days	0	80	124	8.4	(1977)
Chicken	1000	8 weeks	0	n.d.	n.d.	n.d.	Prior et al.
	2000		0	24	41	n.d.	(1980)
			1	n.d.	16	–	
			2	n.d.	n.d.	–	

* Not detected.
† Not determined.

and Bondy 1989). OA is frequently detected in Scandinavian and Balkan countries and occasionally in the USA in commodities such as barley, corn, wheat, oats, rye, peanuts, hay, and green coffee beans (CAST, 1989). The toxin has been related to endemic kidney disease in swine and poultry in Denmark and Sweden. During the course of the swine disease, pigs fed diets containing 200 ppb OA develop pale swollen kidneys characterized by atrophy of the proximal tubules and interstitial cortical fibrosis (Krogh, 1977). These symptoms are similar to those occurring in humans during the course of endemic Balkan kidney disease of the former Yugoslavia, Bulgaria, and Romania. The incidence of this human disease can be correlated to OA content of foods by regions in the Balkan countries (CAST, 1989). OA is detectable in human serum in a number of countries suggesting that there is exposure via food (Kuiper-Goodman and Scott, 1989).

OA accumulates to significant levels in meat and poultry tissue under both experimental and natural conditions. OA accumulation data for several species indicates that the primary sites of residues were the kidney with lesser amounts in liver and muscle (Table 9.3). Krogh et al. (1979) determined that there was no difference in OA residue levels in kidney, muscle, liver, and fat of swine fed 1000 ppb dietary OA for three months or two years. Feed-to-tissue ratios for pig and chicken kidney were less than 100 indicating extensive carryover of OA. Following oral administration, the biological half-life of OA in swine tissues is 4.5 days, indicating that a several week withdrawal period would be needed to eliminate OA residues after exposure to contaminated feed (Krogh et al., 1976; Galtier et al. 1981).

Field surveys of market meat and poultry demonstrate the presence of OA in tissue, primarily kidney (Kuiper-Goodman and Scott, 1989). For example, Josefsson (1979) reported that of 90 abnormal appearing pig kidneys from six Swedish slaughter houses, 22 contained OA at levels greater than 2 ppb with the highest detectable concentration being 88 ppb. Other similar studies in Sweden, Denmark, and former Yugoslavia have reported OA in 25–40% of pig and poultry carcasses that were selected on the basis of abnormal kidneys (Stoloff, 1979).

Losses in total toxin averaged only about 20% when OA-contaminated ground muscle, fat, diced kidneys, and sliced blood pudding were fried at 150–160°C, suggesting that, as in the case of AFB_1, cooking has little effect on OA (Josefsson and Moller, 1980). In view of the potential dangers of having OA in human foods, Danish health authorities have devised a system whereby all suspect pig kidneys are analyzed for the toxin (Van Egmond, 1989). Should OA be detected, the whole carcass is condemned and the source of the contaminated pig feed determined and eliminated. In the USA, regulations concerning the presence of OA in meat or plant products do not exist, due to the apparent low incidence of the toxin in this country.

Sterigmatocystin.   Sterigmatocystin (Figure 9.3), the parent compound of a family of toxins known as the sterigmatocystins, is produced in cultures of *Aspergillus flavus* and *A. versicolor* and *Penicillium luteum* (Busby and Wogan, 1979). Like the aflatoxins, sterigmatocystins are hepatotoxic and hepatocarcinogenic with approximately one-tenth the potency. Although isolated from severely molded wheat, barley, rice, and pecans, sterigmato-cystin is rarely found naturally and thus unlikely to occur as a residue in fresh meats.

Cyclopiazonic acid.   Cyclopiazonic acid (Figure 9.3) is a toxic indoletet-ramic acid produced by species in the *Penicillium* and *Aspergillus* groups. It is a natural contaminant of corn and peanuts and is often coproduced with aflatoxin (Gallagher *et al.*, 1978). Toxic effects include emesis, diarrhoea and convulsions (Morrissey *et al.*, 1985). Norred *et al.* (1985) found that when rats were dosed with cyclopiazonic acid, skeletal muscle tissue contained 48% of the radioactive dose at 6 hours after either intraperitoneal or intragastric administration. After 72 hours, between 3 and 8% of the dose still remained in the muscle. In view of potential human exposure to this mycotoxin via consumption of meat from domestic animals that ingested contaminated feed, there is need for a thorough understanding of both cyclopiazonic acid metabolism and transmission of this mycotoxin into poultry and livestock tissues.

Penicillic acid.   Penicillic acid is a highly reactive unsaturated lactone (Figure 9.3) produced by species of *Aspergillus* and *Penicillium* in corn (Davis and Diener, 1978; Busby and Wogan, 1979). Penicillic acid is moderately toxic to animals and binds directly to sulfhydryl groups or to free amino groups of proteins. Orally administered ^{14}C-labeled penicillic acid is rapidly metabolized and excreted with most residual activity remaining in the blood (Park *et al.*, 1977).

Patulin. Patulin is another reactive unsaturated lactone (Figure 9.3) that is commonly isolated from apples (Davis and Diener, 1978). It is moderately toxic to animals and has been reported to cause stomach lesions. Patulin is rapidly metabolized in rats (Dailey *et al.*, 1975) and since the toxin is not typically found in livestock feeds, is unlikely to accumulate in meat and poultry products.

Citrinin. Citrinin is a moderately toxic compound (Figure 9.3) also produced by *Aspergillus* spp. and *Penicillium* spp. (Davis and Diener, 1978). It has been detected in barley, wheat, and other grains. Citrinin is nephrotoxic and it has been suggested, along with OA, to be involved in swine kidney disease. However, its occurrence is relatively rare and thus probably does not constitute a residue problem in animal tissues.

9.2.2 Fusarium *toxins*

Trichothecenes. Of over 50 known trichothecenes, a group of sesquiterpenoids containing a trichothecene nucleus (Figure 9.4), those most often identified in *Fusarium culture* filtrates are T-2 toxin, diacetoxyscirpenol, and deoxynivalenol (vomitoxin) (Pestka and Casale, 1990). These compounds produce a strong dermatitic reaction indicated by severe local irritation, inflammation, desquamation, and general necrosis (Busby and Wogan, 1979). The biological mode of action for trichothecenes is believed to be via the inhibition of initiation or the elongation–termination steps of protein synthesis. Actively dividing tissues such as bone marrow, lymph nodes, spleen, thymus, testes, ovary, and intestinal mucosa appear most susceptible to these toxins. The potential danger of trichothecenes to human health was first realized in the 1940s during several massive outbreaks (1942–47) of fatal alimentary toxic aleukia in the Soviet Union where overwintered grain containing the toxins was ingested (Joffe, 1978). Sporadic ingestion of sublethal doses of trichothecenes may result in impairment of immune function (Pestka and Bondy, 1990).

Toxinogenic *Fusaria* are often isolated from agricultural commodities such as corn, wheat, barley, and oats (CAST, 1989). Vesonder *et al.* (1973) first isolated the trichothecene, deoxynivalenol, as an emetic principle and feed refusal factor in *Fusarium*-infected corn. The toxin has since been found to occur in the ppm range. Although T-2 toxin has also been isolated in some feed-associated mycotoxicoses, deoxynivalenol appears to be the most common naturally occurring trichothecene.

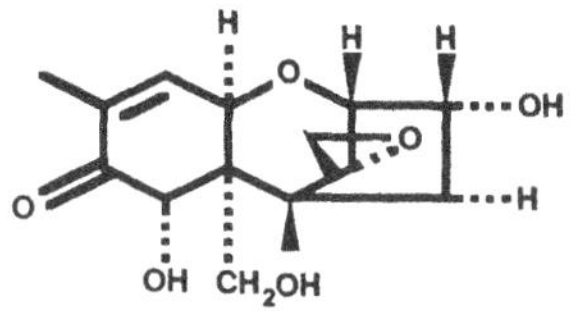

Deoxynivalenol

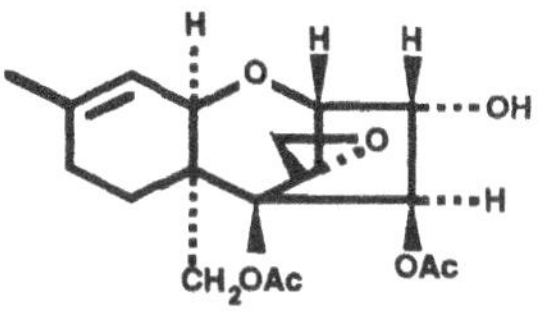

Diacetoxyscirpenol

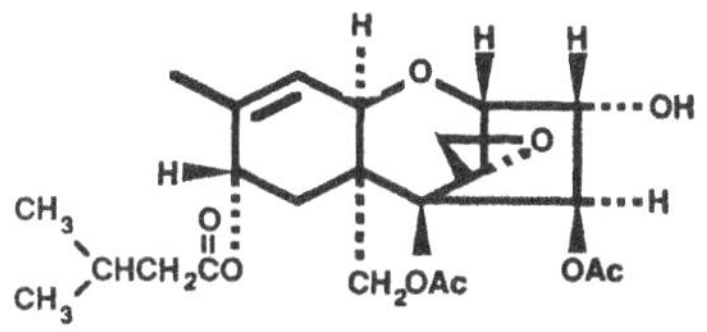

T-2

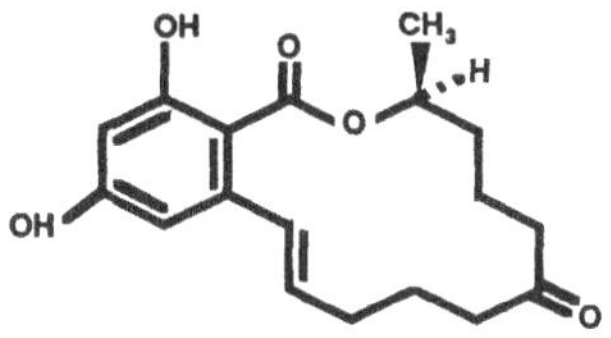

Zearalenone

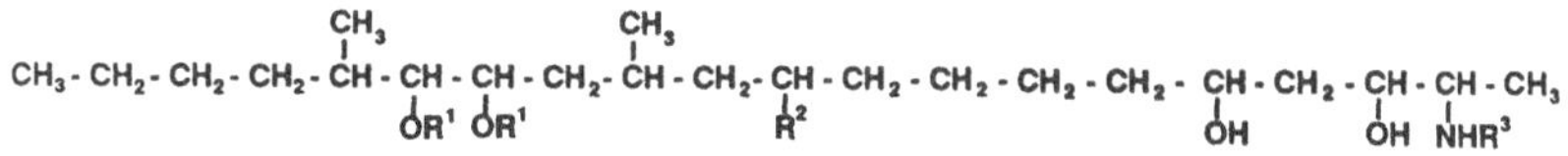

Fumonisins

A_1: R^1 = $COCH_2CH(CO_2H)CH_2CO_2H$; R^2 = OH; R^3 = COCH
A_2: R^1 = $COCH_2CH(CO_2H)CH_2CO_2H$; R^2 = H; R^3 = COCH
B_1: R^1 = $COCH_2CH(CO_2H)CH_2CO_2H$; R^2 = OH; R^3 = H
B_2: R^1 = $COCH_2CH(CO_2H)CH_2CO_2H$; R^2 = R^3 = H

Figure 9.4 Selected *Fusarium* toxins.

Research on metabolism and carryover of trichothecene toxins has dealt primarily with T-2 toxin. Comparison of tissue carryover in liver, muscle, and milk of the cow, chick, and pig (Chi *et al.*, 1978; Robison *et al.*, 1979; Yoshizawa *et al.*, 1981) suggests that liver is the primary site of residue accumulation. More T-2 toxin and T-2 metabolites are transmitted to edible tissue of the chicken than the pig, based on the respective feed to tissue ratios for liver, 41:91, and muscle, 77:500. The lactating cow had higher feed-to-tissue ratios than the other species but comparisons are not possible since the cow study employed a threefold longer withdrawal time. Two of the major metabolites of T-2 toxin found in tissue *in vivo* are toxic and have been identified as a hydroxy derivative and a deacetylated hydroxy derivative of the toxin (Yoshizawa *et al.* 1982). Swanson *et al.* (1982) reported carryover of diacetoxyscirpenol from feed to swine tissues.

Only trace amounts of deoxynivalenol-related residues are transmitted into animal products such as milk (Prelusky *et al.*, 1984), poultry meat and eggs (El-Banna *et al.*, 1983; Kubena *et al.*, 1985; Prelusky *et al.*, 1987, 1989). Contrastingly, pigs absorb deoxynivalenol efficiently but metabolize and eliminate it at a much slower rate than other animal species (Coppock *et al.*, 1985; Prelusky *et al.*, 1988). Nevertheless, recent examination of the tissue of pigs dosed intravenously with deoxynivalenol, suggests that there is not extensive uptake or retention and that accumulation of residues would not occur with prolonged consumption of the toxin (Prelusky and Trenholm, 1991).

Zearalenone.   The zearalenones are a group of resorcyclic lactones, that have estrogenic effects on a number of animal species (CAST, 1989). Zearalenone (Figure 9.4), the parent compound, is often found in temperate regions in corn, wheat, sorghum, barley, and oats. While not acutely toxic, high levels of zearalenone interrupt normal oestrus in sows and cause vulval swelling and enlargement of the uterus. Testicular atrophy and mammary gland hyperplasia develop in young males exposed to zearalenone. Zearalenone affects a number of other species including humans.

Zearalenone is rapidly metabolized in pig, cow, rabbit, rat and humans, and is eliminated in urine and faeces primarily as glucuronide and sulfate conjugates (Mirocha *et al.*, 1981). Oral intubation of rats with ^{3}H-labeled zearalenone does not result in appreciable radioactivity accumulation in liver, kidney, spleen, ovary, brain, or uterus (Mirocha *et al.* 1977). Dairy cows fed diets containing 2 ppm of unlabelled zearalenone for seven weeks did not accumulate the mycotoxin (detection limit 4 ppb) in tissue or milk (Shreeve *et al.*, 1979). In contrast, Mirocha *et al.* (1981) found that when cows were fed diets containing 25 ppm zearalenone for 7 days, a feed-to-tissue ratio (zearalenone to free zearalenone plus zearalenone conjugates) of 17 resulted in milk, indicating a very high potential for carryover. Further clarification of the potential for carryover of zearalenone in edible tissue is likely with improved analytical procedures.

Fumonisins.   The fumonisins (Figure 9.4) are a newly isolated group of mycotoxins produced by *Fusarium moniliforme* strains that were associated with equine leukoencephalomacia (ELEM). Fumonisin $B_1(FB_1)$ experimentally induces ELEM (Kellerman *et al.*, 1990) and porcine pulmonary edema (Harrison *et al.*, 1990), causes hepatic cancer in rats and exhibits cancer-promoting activity (Gelderblom *et al.*, 1988). Very little is known of the mechanisms of toxicity or potential for carryover of this important group of toxins into edible animal tissue.

9.3 Fungal growth and mycotoxinogenesis on meats

Critical parameters which affect the growth of bacteria and fungi on a rich nutrient source such as meat include temperature, water activity (a_w), pH, and oxidation–reduction potential (E_h). Even though fungi tolerate wide variations in these parameters, they are dominated by bacteria in meats because of their relatively slow growth rates and high E_h requirements (Jay, 1978). For a mold to grow on meat, ample oxygen must usually be present and some other intrinsic or extrinsic factors modified to inhibit dominant bacterial groups and allow the slower-growing fungi to colonize the substrate. Specifically, the ripening and storage conditions used for cured and aged meats often facilitate the growth of several genera of fungi. In meat products where fungal growth is considered spoilage, it can be prevented by light smoking, water activity control, vacuum packaging and treatment with sorbate and pimaricin (Leistner, 1984). However, in other products, such as Hungarian- and Italian-type fermented sausages and country-cured hams, molds are desirable as part of a 'natural' fermentation process (Leistner and Ayres 1968). These contribute to the aesthetic appearance of mold on these special cured meats that is traditionally expected by the consumer. The molds also contribute to improved flavour of fermented meats via inhibition of fat oxidation, proteolytic and lipolytic changes, or inhibition of spoilage bacteria. Molded sausages may dry in a slow uniform manner resulting in improved quality and reduced weight loss.

9.3.1 Characteristic mycoflora of mold-ripened meats

Mold-fermented raw sausages (salami) are of primary importance in Southern and South-eastern Europe (Leistner, 1984). This type of salami is prepared by encasing ground beef and pork mixed with spices and curing agents and subjecting these to ripening in which the relative humidity is decreased from 95 to 75%. A uniform white-grey mold appears during a ripening period, lasting days to weeks. Hungarian-type sausages are smoked lightly during ripening, whereas most other types are not smoked. Fungal growth usually occurs during the first five days of ripening and is dictated by the characteristic microbial flora of the ripening chamber (Ciegler et al., 1972). Penicillia make up the primary indigenous flora of salami-ripening rooms, about 70–80% of which are potential toxin producers (Leistner and Pitt, 1977; Eckhardt et al., 1979; Leistner and Eckhardt, 1979b).

Unsmoked raw cured hams that also characteristically exhibit mold growth include US country-cured ham, Yugoslave kraski prsut, Italian speck and Swiss Bunderfleisch (Leistner, 1984). For example, in the production of country-cured hams, the hams are dry-cured, with or without smoking, and aged for between 6 and 24 months (Bullerman *et al.*, 1969b). Ageing is usually done in an environment such as a room or attic where humidity and temperature are not regulated. The resulting ham often has extensive fungal growth, usually green, covering the flesh. Because of the absence of a casing, hams present more of a risk of undesirable molds than salami, with most mycotoxin production occurring in the first 15 mm below the surface (Leistner, 1984).

Sampling cured meats from 32 different European and US producers, Leistner and Ayres (1968) isolated 307 fungal strains from fermented sausages and country-cured ham. In the fermented sausages, the authors found *Penicillium*, *Scopulariopsis* and *Aspergillus* in 89, 41, and 33%, respectively, of the samples examined. Minor flora of the fermented sausages included *Rhizopus* (11%), *Mucor* (4%), and *Mortierella* (4%). *Aspergillus* and *Penicillium* predominated in the country-cured hams, and were found in 90 and 83% of the samples, respectively. Other genera isolated from the ham samples were *Cladosporium* (30%), *Rhizopus* (13%), *Alternaria* (13%), *Scopulariopsis* (8%), *Paecilomyces* (8%), and *Oospora* (8%). The authors observed that country-cured hams wrapped in paper and held at ambient humidity and temperature during the ripening process were heavily molded on the flesh portions but not on the skin or fat. Fatty portions of the wrapped hams exhibited only moderate oxidization even after 9–12 months. Unwrapped hams held at low relative humidity (65%) during ripening had little mold growth but their fat became strongly oxidized after several months. Aspergilli generally grew better in the low a_w country-cured hams than in the higher a_w fermented sausages where the penicillia were predominant.

In a survey of German, Italian and Hungarian salami, Leistner and Eckhardt (1979a) determined that 21, 66 and 77% of the *Penicillium* isolates, respectively, were capable of producing mycotoxins; ochratoxin A and cyclopiazonic acid were the predominant mycotoxins.

Sutic *et al.* (1972) isolated 562 fungi from 356 country-cured hams. Of these, 403 were *Penicillium*, 121 were *Aspergillus*, and the remainder were *Cladosporium*, *Alternaria*, and other genera. Again, aspergilli were most often isolated from hams stored under dry conditions, whereas penicillia were associated with humid storage conditions. As was found in the previous study, members of the *A. glaucus* group predominated among the aspergilli. In contrast to country-cured hams, penicillia are predominantly isolated from the surface of *speck* and *Bundnerfleisch* (Leistner and Pitt, 1977; Leistner *et al.*, 1981)

9.3.2 Mycotoxinogenesis

Experimental inoculation of meat with toxinogenic fungi can result in mycotoxin production (Hoffmann *et al.*, 1981; Leistner *et al.*, 1981; Pestka,

1986). Aflatoxins may appear in meat products as a direct result of the growth of toxinogenic aspergilli. Bullerman *et al.* (1969a) found that fresh beef, bacon, and ham could support both the growth of inoculated *A. flavus* and *A. parasiticus* as well as aflatoxin production. Storage at 20°C for 14 days was optimal for toxin production with levels in beef exceeding 600 ppb. However, meats stored at 10°C became spoiled by bacteria and yeast before significant *Aspergillus* growth. Thus, products less susceptible to bacterial spoilage, such as bacon and ham, can potentially support aflatoxin production when subjected to ample oxygen and temperature abuse. It is, therefore, good practice for the consumer to discard products of this type when heavy mold contamination is apparent on them.

The fermentation conditions used for country-cured hams and aged salamis hypothetically make these meat products especially susceptible to colonization by toxinogenic aspergilli. Aflatoxins are elaborated when Italian-type and Hungarian-type salamis, summer sausages, and country-cured hams are inoculated with *A. flavus* or *A. parasiticus* before ageing (Bullerman *et al.*, 1969b; Alvarez-Barrera *et al.*, 1983a). Toxin production is greatest at temperatures near 30°C and relative humidities greater than 85% but can be inhibited by aging the meats at temperatures between 10 and 15°C and relative humidities below 80%. Smoking and curing ingredients also tend to inhibit aflatoxin elaboration (Bullerman *et al.*, 1969b; Meier and Marth 1977). It is important to note that *A. flavus* and *A. parasiticus* are rarely encountered in fermented meat products. For example, of 562 fungi isolated from country-cured hams, only three were *A. flavus* (Sutic *et al.*, 1972). Leistner and Ayres (1968) detected only one strain of *A. flavus* from the 307 mold strains isolated from country-cured hams and European-type salamis. Even if spores of toxinogenic aspergilli are present on a meat, they must successfully compete with other molds in order to colonize and produce aflatoxins. Up to this time, there were no reports of aflatoxins being detected in fermented meat products taken from the market shelf. Thus, the danger of human aflatoxin exposure by this route is probably minimal.

Under ideal growth conditions, aspergilli fungi can also elaborate sterigmatocystin in aged, cured meats. Halls and Ayres (1973) found that 10 of 16 isolates of *Aspergillus versicolor* from country-cured hams were capable of producing sterigmatocystin. When three of these were inoculated onto sterile ham slices, they were shown to produce the toxin at levels as high as 800 ppb. Sterigmatocystin has not been detected in market hams or sausages, although its presence has not been systematically investigated.

Both *A. ochraceus* and *Penicillium viridicatum*, species which often produce ochratoxin A, have been isolated from country-cured hams (Leistner and Ayres, 1968; Strzelecki *et al.*, 1969; Sutic *et al.*, 1972). Escher *et al.* (1973) found that both of two strains of *A. ochraceus* isolated from country-cured hams produced OA and ochratoxin B in culture and in hams; however, none of six *P. viridicatum* strains yielded octratoxin A or B when cultured on ham. Optimum conditions for toxin production were 25–30 and a total moisture content of 45%. After 21 days' incubation, one-third of the toxin was found in mycelia on the ham surface while the remainder penetrated the meat to a distance of 0.5 cm. Systematic market studies of ochratoxin occurrence in marketplace-aged cured hams and sausage have yet to be conducted.

Other *Penicillium* toxins are potentially important in mold-ripened meats. Leistner (1984) reported that cyclopiazonic acid is the major mycotoxin produced by penicillia isolated from mold-fermented salamis suggesting that these meat products should be surveyed for this mycotoxin. Ciegler *et al.* (1972) also determined that 10% of 346 *Penicillium* cultures isolated from mold-fermented salamis produced penicillic acid in liquid media. When five of these mold strains were inoculated onto sausages and ripened for 70 days, toxin was not detected. It was hypothesized that if penicillic acid was formed during the ageing of meats, it would react with amino acids in the meats and be rendered non-toxic. Finally, under experimental conditions, *Penicillium viridicatum* can produce citrinin in county-cured hams (Wu *et al.*, 1974a,b)

9.3.3   Mold starter cultures

In view of the established potential for toxinogenic mold occurrence and mycotoxinogenesis in mold-ripened meats produced under traditional conditions, it would be prudent to move toward more carefully controlled fermentations that employ starter cultures. The required technological approaches as described by Leistner *et al.* (1989) are to: (i) isolate the desired mold from high-quality meat products; (ii) demonstrate suitability for competitive growth on the product and for acceptable appearance and flavour; and (iii) develop diverse strains for specific products by classical isolation from natural reservoirs and by genetic engineering. Absence of toxicity must be verified by screening and confirmatory method.

A suitable candidate for starter cultures is *Penicillium nalgiovense* which is typically found on mold-fermented sausages and hams produced in Europe (Leistner *et al.*, 1989). This species produces white-grey growth and contributes to the desired sensory qualities of these products. Mintlaff and Leistner (1972) first isolated a strain of *P. nalgiovense* (Sp290) that was suitable for use as a starter culture and has been used commercially under the name 'Edelschimmel Kulmbach'. Several other *Penicillium* starters have been offered by French companies (Leistner *et al.* 1989).

To expand available mold starter strains, Leistner *et al.* (1989) have screened 119 *P. nalgiovense* isolates and found nine acceptable based on morphology, toxicity screening and product quality attributes. After further toxicologic evaluation, one strain (Sp 1372) was selected for improvement by genetic engineering. Methods have been devised for protoplast production and transformation of *P. nalgiovense* that should eventually enable introduction of genes which improve sensory and altagonistic functions of the starter cultures (Geisen and Leistner, 1989; Geisen *et al.*, (1989).

9.4 Summary and conclusions

Trace levels of mycotoxins and their metabolites have been experimentally shown to carry over from feeds to the edible tissues of food-producing animals. Generally, the degree of mycotoxin transmission is related to total dose level rather than duration of dose. To date, there is no evidence to suggest that the levels of transmitted mycotoxin pose a threat of acute toxicity but there is potential for long-term effects such as carcinogenesis or immunotoxicity.

Carryover of aflatoxins has been most thoroughly studied and is subject to strict regulation at the feed level because of the ability of these toxins to act as carcinogens. Ochratoxin A residues in meat may, however, pose a significant health problem in certain European countries and are therefore monitored by regulatory authorities. Although there are sporadic incidences of *Fusarium* toxins, particularly deoxynivalenol and zearalenone, in cereal grains, risks from carryover into edible tissue appear minimal. Suitable information regarding feed to tissue carryover of cyclopiazonic acid and fumonisin is lacking. It is not yet known whether these compounds impair human health when present in the diet. Since cyclopiazonic acid and fumonisin are likely to be present in the diet of food-producing animals, further data on their potential to appear as residues need to be developed.

Relatively little is known of the actual levels of mycotoxins and their residues in market meat. There is thus a need for systematic assessment of this possibility, perhaps using immunochemical assay as a screening procedure. Analytical enzyme immunoassays have been devised for mycotoxins that allow rapid and specific quantitation of the toxins at the picogram level without the cleanup problems encountered by other methods (Pestka, 1988). Although these have been applied primarily to plant products and milk, they might similarly be used to screen for the presence of mycotoxins in meats. One approach might involve the use of mycotoxin plasma level as a diagnostic 'indicator' of tissue contamination. Development of a comprehensive database on mycotoxin tissue residues in market-ready meats would greatly enhance our ability to make rational judgments on the risks and benefits of using feed contaminated with mycotoxins at low-to-moderate levels.

Fungi provide quality attributes to certain aged and cured meats. These are primarily esthetic, although improved flavor, texture, and preservation characteristics also result from growth of desirable molds. From a health standpoint, fungal species, particularly penicillia, isolated from fermented meats are non-pathogenic but many can produce mycotoxins in meat under appropriate environmental conditions. As with transmitted residues, there is scant evidence to suggest that market ready mold-ripened sausages and hams contain dangerous levels of mycotoxins. Nevertheless, systematic market surveys for mycotoxins of fermented meat products could clarify this issue.

The traditional means for 'inoculating' a fermented sausage is simply by hanging it in a room that has been used for aging and molding of sausages for many years and thus contains high concentrations of desirable mold spores. In order to eliminate the potential mycotoxin hazard, it might be preferable to utilize fungi isolated from fermented meats as starter cultures for fermented sausages and hams. Major considerations in selection of starter strains are their potential for pathogenicity, antibiotic production, and toxinogenicity. *Penicillium nalgiovense* has already been used success-fully towards this end and there has been progress made towards improving this strain by genetic engineering. Further studies, similar to those conducted on the lactic acid bacteria and micrococci, will be helpful in establishing the efficacy of using newly developed mold starter cultures in aged, fermented meats.

References

Alvarez-Barrera, V., Pearson, A.M., Price, J.F., Gray, J.I. and Aust, S.D. (1983) Some factors influencing aflatoxin production in fermented sausages. *J. Food Sci.*, **47**, 1773–1775.

Austwick, P.K.C. (1978) Aflatoxicosis in poultry. In *Mycotoxic Fungi, Mycotoxins, Mycotoxicoses. An Encyclopedic Handbook*. Eds T.D. Wyllie and L.G. Morehouse, pp. 279–331. Marcel Dekker, New York.

Bullerman, L.B., Hartman, P.A. and Ayres, J.C. (1969a) Aflatoxin production in meats. I. Stored meats. *Appl. Microbiol.*, **18**, 714–717.

Bullerman, L.B., Hartman, P.A. and Ayres, P.A. (1969b). Aflatoxin production in meats. II. Aged dry salamis and aged country cured hams. *Appl. Microbiol.*, **18**, 718–722.

Busby, Jr., W.F. and Wogan, G.F. (1979) Mycotoxins and mycotoxicoses. In *Food-Borne Infections and Intoxications*, 2nd Ed. Eds H. Riemann and F.L. Bryan, pp. 519–610. Academic Press, New York.

Carlton, W.W. and Szczech, G.M. (1978) Mycotoxicoses of laboratory animals. In *Mycotoxic Fungi, Mycotoxins, Mycotoxicoses. An Encyclopedic Handbook*. Eds T.D. Wyllie and L.G. Morehouse, pp. 333–487. Marcel Dekker, New York.

CAST (1989) *Mycotoxins: Economic and Health Risks*. Report No. 116. Council for Agricultural Science and Technology, Ames

Chen, C., Pearson, A.M., Coleman, T.H., Gray, J.I., Pestka, J.J. and Aust, S.D. (1984) Tissue deposition and clearance of aflatoxins from broiler chicken fed a contaminated diet. *Food Chem. Toxicol.*, **22**, 447–451.

Chi, M.S., Robison, T.S., Mirocha, C.J., Behrens, J.C. and Shimoda, W. (1978) Transmission of radioactivity into eggs from laying hens administered tritium labeled T-2 toxin. *Poult. Sci.*, **57**, 1234–1239.

Ciegler, A., Mintzlaff, A-J., Weisleder, D. and Leistner, L. (1972) Potential production and detoxification of penicillic acid in mold-fermented sausage (salami). *Appl. Microbiol.*, **24**, 114–119.

Coppock, R.W., Swanson, S.P., Gelberg, H.B., Koritz, G.D., Hoffman, W.E., Buck, W.B. and Vesonder, R.F. (1985) Preliminary study of the pharmacokinetics and toxicopathy of deoxynivalenol (vomitoxin) in swine. *Am. J. Vet. Res.*, **46**, 169–174.

Cysewski, S.J., Pier, S.J., Engstrom, G.W., Richard, J.L., Dougherty, R.W. and Thurston, J.R. (1968) Clinical features of acute aflatoxicosis in swine. *Am. J. Vet. Res.*, **29**, 1577–1590.

Dailey, R.E., Blaschka, A.M. and Brouwer, E.A. (1975) Absorption, distribution, and excretion of [^{14}C]patulin by rats. *J. Toxicol. Environ. Health*, **3**, 479–485.

Davis, N.D. and Deiner, U.L. (1987) Mycotoxin. In *Food and Beverage Mycology*. 2nd Edn. Ed. L.R. Beuchat, pp. 517–570. AVT Publishing, Westport, CT.

Eckardt, C., Ramming, G., Trapper, D. and Leistner, L. (1979) Vorkommen toxinogener *Penicillium*-Arten bei Lebens- und Futtermitteln. *Jahresber. Bundesanst. Fleischforsch. Kulmbach C*, pp. 24–25.

El-Banna, A.A., Hamilton, R.M.G., Scott, P.M. and Trenholm, H.L. (1983) Nontransmission of deoxynivalenol (vomitoxin) to eggs and meat in chickens fed deoxynivalenol contaminated diets. *J. Agric. Food Chem.*, **31**, 1381–1384.

Escher, F.E., Koehler, P.E. and Ayres, J.C. (1973) Production of ochratoxins A and B on country cured ham. *Appl. Microbiol.*, **26**, 27–30.

Essigmann, J.M., Croy, R.G., Bennett, R.A. and Wogan, G.N. (1982) Metabolic activation of aflatoxin B: Patterns of DNA adduct formation, removal, and excretion in relation to carcinogenesis. *Drug Metab. Rev.*, **13**, 581–602.

Frye, C.E. and Chu, F.S. (1977) Distribution of ochratoxin A in chicken tissues and eggs. *J. Food Safety*, **1**, 147–159.

Furtado, R.M., Pearson, A.M., Gray, J.I., Hogberg, M.G. and Miller, E.R. (1981) Effects of cooking and/or processing upon levels of aflatoxin in meat from pigs fed a contaminated diet. *J. Food Sci.*, **46**, 1306–1308.

Furtado, R.M., Pearson, A.M., Hogberg, M.G., Miller, E.R., Gray, J.I. and Aust, S.D. (1982) Withdrawal time required for clearance of aflatoxins from pig tissues. *J. Agric. Food Chem.*, **30**, 101–106.

Gallagher, R.T., Richard, J.L., Stahr, H.M. and Cole, R.J. (1978) Cyclopiazonic acid production by aflatoxinogenic and non-aflatoxinogenic strains of *Aspergillus flavus*. *Mycopathologia*, **66**, 31–36.

Galtier, P., Alvinerie, M. and Charpenteace, J.L. (1981) The pharmacokinetic profiles of ochratoxin A in pigs, rabbits and chickens. *Food Cosmet. Toxicol.*, **19**, 735–738.

Geisen, R. and Leistner, L. (1989) Transformation of *Penicillium nalgiovense* with the S and gene of *Aspergillus nidulans*. *Curr. Genet.*, **15**, 307–309.

Geisen, R., Glenn, E. and Leistner, L. (1989) Production and regeneration of protoplasts from *Penicillium nalgiovense*. *Lett. Appl. Microbiol.*, **8**, 99–100.

Gelderblom, W.C.A., Jaskiewicz, K., Marasas, W.F.O., Thiel, P.G. Horak, R.M., Vegglaar, R. and Kriek, N.P.J. (1988) Fumonisins — novel mycotoxins with cancer promoting activity produced by *Fusarium moniliform*. *Appl. Environ. Microbiol.*, **54**, 1806–1811.

Gregory, J.F. III, Goldstein, S.L. and Edds, G.T. (1983) Metabolite distribution and rate of residue clearance in turkeys fed a diet containing aflatoxin B_1. *Food. Chem. Toxicol.*, **21**, 463–467.

Halls, N.A. and Ayres, J.C. (1973) Potential production of sterigmatocystin on country cured ham. *Appl. Microbiol.*, **26**, 636–637.

Harrison, L.R., Colvin, B.M., Greene, J.T., Newman, L.E. and Cole, J.R. (1990) Pulmonary edema and hydrothorax in swine produced by fumonisin B_1, a toxic metabolite of *Fusarium moniliforme*. *J. Vet. Diagn. Invest.*, **2**, 217–221.

Hoffmann, G., Leistner, L. and Trapper, D. (1981) Mykotoxinbildung in Rohschinken. *Mitteil. ungsbl. Bundesanst. Fleischforsch. Kulmbach N*. 71, 4444–4446.

Jacobson, W.C., Harmeyer, W.C., Jackson, J.E., Armbrecht, B. and Wiseman, A.G. (1978) Transmission of aflatoxin B_1 into tissues of growing pigs. *Bull. Environ. Contam. Toxicol.*, **19**, 156–161.

Jaggi, W., Lutz, W.K., Luthy, J., Zweifel, U. and Schlatter, C. (1980) *In vivo* covalent binding of aflatoxin metabolites isolated from animal tissue to rat-liver DNA. *Food Cosmet. Toxicol.*, **18**, 257–260.

Jay, J.M. (1978). Meats, Poultry, and seafoods. In *Food and Beverage Mycology*. Ed. L.R. Beuchat, p. 129. Avi Publishing, Westport, CT.

Joffe, A.Z. (1978) *Fusarium poae* and *F. sporotrichoides* as principal causal agents of alimentary toxic aleukia. In *Mycotoxic Fungi, Mycotoxins. Mycotoxicoses. An Encyclopedia Handbook*. Eds T.D. Wyllie and L.G. Morehouse, p. 21–42 Marcel Dekker, New York.

Josefsson, E. (1979) Study of ochratoxin A in pig kidneys. *Var Foeda*, **31**, 415–418.

Josefsson, B.G.E. and Moller, T.E. (1980) Heat stability of ochratoxin A in pig products. *J. Sci. Food Agric.*, **31**, 1313–1319.

Kellerman, T.S., Marasas, W.F.O., Thiel, P.G., Gelderblom, W.C.A., Cawood, M. and Coetzer, J.A.W. (1990) Leukoencephalomalacia in two horses induced by oral dosing of fumonisin B_1. *Onderstepoort. J. Vet. Res.*, **57**, 269–275.

Krogh, P. (1977) Ochratoxins. Rodricks, J.V., Hesseltine, C.W. and Mehlmann, M.A. In *Mycotoxins in Human and Animal Health*, pp. 489–506 Pathotox Publishers, Park Forest South, IL.

Krough, P., Elling, F., Hald, B., Larsen, A.E., Lillehoj, E.B., Madsen, A. and Mortensen, H.P. (1976) Time-dependent disappearance of ochratoxin A residues in tissues of bacon pigs. *Toxicology*, **6**, 235–240.

Krogh, P., Elling, F., Friis, C., Hald, B., Larsen, A.E., Lillehoj, E.B., Madsen, A., Mortensen, H.P., Rasmussen, F. and Raunshov, U. (1979) Porcine nephropathy induced by long-term ingestion of ochratoxin A. *Vet. Pathol.*, **16**, 446–452.

Kubena, L.F., Swanson, S.P., Harvey, R.B., Fletcher, O.J., Rowe, L.D. and Phillips, T.D. (1985) Effects of feeding deoxynivalenol (DON, vomitoxin) contaminated wheat to growing chicks. *Poult. Sci.*, **64**, 1649–1655.

Kuiper-Goodman, T. and Scott, P.M. (1989) Risk assessment of the mycotoxin ochratoxin A. *Biomed. Environ. Sci.*, **2**, 179–248.

Leistner, L. (1984) Toxinogenic penicillia occurring in feeds and foods: A review. *Food Tech. Aust.* **36**, 404–406.

Leistner, L. and Ayres, J.C. (1968) Molds and meats. *Fleischwirtschaft*, **48**, 62–65.

Leistner, L. and Pitt, J.I. (1977) Miscellaneous *Penicillium* toxins. In *Mycotoxins in Human and Animal Health*. Eds J.V. Rodricks, C.W. Hesseltine and M.A. Mehlmann, pp. 639–653. Park Forest South IL.

Leistner, L. and Eckardt, C. (1979a) Vorkommen toxinogener Pinicillien bei Fleischerzeugnissen. *Proc. 25th Eur. Meeting Meat Res. Workers, Budapest*, **2**, pp. 485–491.

Leistner, L. and Eckardt, C. (1979b) Vorkommen toxinogener Penicillien bei Fleischerzeugnissen. *Fleischwirtschaft*, **59**, 1982–1986.

Leistner, L. and Eckardt (1981) *C. Schimmelpilze und Mykotoxine in Fleisch und Fleischezeugnissen. Mykotoxine in Lebensmitteln*. Ed. J. Reis, pp. 297–341. Gustav Fischer, Stuttgart.

Leistner, L., Geisen, R. and Fink-Gremmels, J. (1989) Mould-fermented foods of Europe: hazards and developments. Eds S. Natori, K. Hashimoto and Y. Ueno. pp. 145–154. In *Mycotoxins and Phycotoxins*, '88. A collection of papers presented at the Seventh International IUPAC Symposium on Mycotoxins and Phycotoxins, Tokyo, Japan.

Luthy, J., Zweifel, U. and Schlatter, C.H. (1980) Metabolism and tissue distribution of [^{14}C] aflatoxin B_1 in pigs. *Food Cosmet. Toxicol.*, **18**, 253–256.

Meier, K.R. and Marth, E.H. (1977) Production of aflatoxin by *Aspergillus parasiticus* NRRL-2999 during growth in the presence of curing salts. *Mycopathologia*, **61**, 77–83.

Miller, D.M., Wilson, D.M., Wyatt, R.D., McKinney, J.K., Cromwell, W.A. and Stuart, B.P. (1982) High performance liquid chromatographic determination and clearance time of aflatoxin residues in swine tissues. *J. Assoc. Off. Anal. Chem.*, **65**, 1–7.

Mintzlaff, H-J. and Leistner, L. (1972) Untersuchungen zur Selektion eines technologisch geeigneten und toxikologisch unbedenklichen Schimmelpilz-Stammes für die Rohwurst-Herstelling, ZBl. *Vet. Med. B*, **19**, 291–300.

Mirocha, C.J., Pathre, W.V. and Christensen, C.M. (1977) Zearalenone. In *Mycotoxins in Human and Animal Health*. Eds. J.V. Rodricks, C.W. Hesseltine and M.A. Mehlman, pp. 345–364. Pathotox Publishers, Park Forest South, IL.

Mirocha, C.J., Pathre, S.V. and Robinson, T.S. (1981) Comparative metabolism of zearalenone and transmission into bovine milk. *Food Cosmet. Toxicol.*, **19**, 25–30.

Morrissey, R.E., Norred, W.P., Cole, J.R. and Dorner, J. (1984) Toxicity of the mycotoxin, cyclopiazonic acid to Sprague–Dawley rats. *Toxic. Appl. Pharmacol.*, **77**, 94–107.

Moss, E.J., Judah, D.J., Przybylski, M. and Neal, G. (1983) Some mass-spectral and NMR analytical studies of a glutathione conjugate of aflatoxin B_1. *Biochem. J.*, **210**, 227.

Neal, G.E. and Colley, P.J. (1979) The formation of 2,3-dihydro-2,3-dihydroxy aflatoxin B_1 *in vitro* by rat liver microsomes. *FEBS Lett.*, **101**, 382–386.

Neal, G.E., Judah, D.J., Stripe, F. and Patterson, D.S. (1981). The formation of 2,3-dihydroxy-2,3-dihydro-aflatoxin B_1 by the metabolism of aflatoxin B_1 by liver microsomes isolated from certain avian and mammalian species and the possible role of this metabolite in the acute toxicity of aflatoxin B_1. *Toxicol. Appl. Pharmacol.*, **58**, 431–438.

Neff, G.L. and Edds, G.L. (1981) Aflatoxin B_1 and M_1: Tissue residues and feed withdrawal profile in young growing pigs. *Food Cosmet. Toxicol.*, **19**, 739–742.

Nixon, J.E., Hendricks, J.D., Pawloswki, N.E., Loveland, P.M. and Sinnhuber, R.O. (1981) Carcinogenicity of aflatoxicol in Fischer 344 rats. *J. Nat. Cancer Inst.*, **66**, 1159–1163.

Norred, W.P., Morrisey, R.E., Riley, R.T., Cole, R.J. and Dorner, W.J. (1985) Distribution, excretion and skeletal muscle effects of the mycotoxin [^{14}C] cyclopiazonic acid in rats. *Food, Chem. Toxic.*, **23**, 1069–1076.

Park, D.L. and Rua, S.M. Jr. (1990) Biological evaluation of aflatoxins and metabolites in animal tissues. *Drug Metab. Rev*, **22**(6–8), 871–890.

Park, D.L., Dailey, R.E., Friedman, L. and Heath, J.L. (1977) The absorption, distribution, and excretion of ^{14}C-penicillic acid by rats. *Ann. Nutr. Aliment.*, **31**, 919–934.

Patterson, D.S.P., Shreeve, B.J., Roberts, B.A., Berrett, S., Brush, P.J. and Glancy, E.M. (1981) Effect on calves of barley naturally contaminated with ochratoxin A and groundnut meal contaminated with low concentrations of aflatoxin B_1. *Res. Vet. Sci.*, **31**, 213–218.

Pestka, J.J. (1986) Fungi and mycotoxins in meats. *Adv. Meat Res.* **2**, 277–309.

Pestka, J.J. (1988) Enhanced surveillance of foodborne mycotoxins by immunochemical assay. *J. Assoc. Off. Anal. Chem.*, **71**, 1075–1081.

Pestka, J.J. and Bondy, G.S. (1989) Alteration of immune function following dietary mycotoxin exposure. *Can. J. Physiol. Pharmacol.*, **68**, 1009–1016.

Pestka, J.J. and Casale, W.L. (1990) Naturally occurring fungal toxins. In *Food Contamination from Environmental Sources*. Ed. J.O. Nriagu and M.S. Simmons, pp. 613–638.

Phillips, T.D., Kubena, L.F., Harvey, R.B., Taylor, D.R. and Heidelbaugh, N.D. (1988) Hydrated sodium calcium aluminosilicate: A high affinity sorbent for aflatoxin. *Poult. Sci.*, **67**, 243–247.

Prelusky, D.B., Hamilton, R.M.G. and Trenholm, H.L. (1989) Transmission of residues to eggs following long-term administration of ^{14}C-labelled deoxynivalenol to laying hens. *Poult. Sci.*, **68**, 744–748.

Prelusky, D.B., Hartin, K.E., Trenholm, H.L. and Miller, J.D. (1988) Pharmacokinetic fate of ^{4}C-labelled deoxynivalenol in swine. *Fund. Appl. Toxicol.*, **10**, 276–286.

Prelusky, D.B. and Trenholm, L. (1991) Tissue distribution of deoxynivalenol in swine dosed intravenously. *J. Agric. Food Chem.* **39**, 748–751.

Prelusky, D.B., Trenholm, H.L., Lawrence, G.A. and Scott, P.M. (1984) Nontransmission of deoxynivalenol (vomitoxin) to milk following oral administration to dairy cows. *J. Environ. Sci. Health*, **B19**, 593–609.

Prelusky, D.B., Trenholm, H.L., Hamilton, R.M.G. and Miller, J.D. (1987) Transmission of ^{4}C-deoxynivalenol to eggs following oral administration to laying hens. *J. Agric. Food Chem.*, **35**, 182–186.

Prior, M.G., O'Neil, J.B. and Sisodia, C.S. (1980) Effects of ochratoxin A on growth response and residues in broilers. *Poult. Sci.*, **59**, 1254–1258.

Richard, J.L., Pier, A.C., Stubblefield, R.D., Shotwell, O.L., Lyon, R.L. and Cutlip, R.C. (1983) Effect of feeding corn naturally contaminated with aflatoxin on feed efficiency, physiological, immunological, and pathological changes and tissue residues in steers. *Am. J. Vet. Res.*, **44**, 1294–1299.

Robison, T.S., Mirocha, C.J., Kurtz, H.J., Behrens, J.C., Chi, M.S., Weaver, G.A. and Nystrom, S.D. (1979) Transmission of T-2 toxin into bovine and porcine milk. *J. Dairy Sci.*, **62**, 637.

Rodricks, J.V. and Stoloff, L. (1977) Aflatoxin residues from contaminated feed in edible tissues of food-producing animals. In *Mycotoxins in Human and Animal Health*. Eds J.V. Rodricks, C.W. Hesseltine and M.A. Mehlman, pp. 67–79. Pathotox Publishers, Park Forest South, IL.

Schoenhard, G.L., Hendricks, J.D., Nixon, J.E., Lee, D.J., Wales, J.H., Sinnhuber, R.O. and Pawlowski, N.E. (1981) Aflatoxicol-induced hepatocellular carcinoma in rainbow trout and the synergistic effects of cyclopropenoid fatty acids. *Cancer Res.*, **41**, 1011–1014.

Shank, R.C. (1978) Mycotoxicoses of man: Dietary and epidemiological conditions. In *Mycotoxic Fungi, Mycotoxins, Mycotoxicoses, An Encyclopedic Handook.* Eds T.D. Wyllie and L.G. Morehouse, Vol. 3, p. 1. Marcel Dekker Inc., New York, USA.

Shreeve, B.J., Patterson, D.S.P. and Roberts, B.A. (1979) The 'carry-over' of aflatoxin, ochratoxin, and zearalenone from naturally contaminated feed to tissues, urine and milk of diary cows. *Food Cosmet. Toxicol.*, **17**, 51–152.

Sinnhuber, R.O., Wales, J.H., Engebrecht, R.H., Amend, D.F., Kray, W.D., Ayres, J.L. and Ashton, W.E. (1966). Aflatoxins in cottonseed meal and hepatoma in rainbow trout. *Fed. Proc.*, **24**, 627.

Stoloff, L. (1979) Mycotoxin residues in edible animal tissues. In *Interactions of Mycotoxins in Animal Production*, p. 157. NAS, Washington, DC, USA.

Stoloff, L. (1980) Aflatoxin M_1 in perspective. *J. Food Protect.*, **43**, 226–230.

Strzelecki, E., Lillard, H.S. and Ayres, J.C. (1969) Country cured ham as a possible source of aflatoxin. *Appl. Microbiol.*, **18**, 938–939.

Stubblefield, R.D., Pier, A.C., Richard, J.S. and Shotwell, O.L. (1983) Fate of aflatoxins in tissues, fluids, and excrements from cows dosed orally with aflatoxin B_1. *Am. J. Vet. Res.*, **44**, 1750–1752.

Sutic, M.J., Ayres, J.C. and Koehler, P.E. (1972) Identification and aflatoxin production of molds isolated from country cured hams. *Appl. Microbiol.*, **23**, 656–658.

Trucksess, M.W., Stoloff, L., Brumely, W.C., Wilson, D.M., Hale, O.M., Sangster, L.T. and Miller, D.M. (1982) Aflatoxicol and aflatoxins B_1 and M_1 in tissues of pigs receiving aflatoxin. *J. Assoc. Off. Anal. Chem.*, **65**, 884–887.

Trucksess, M.W., Richard, J.C., Stoloff, L., McDonald, S. and Brumely, W.C. (1983a) Absorption and distribution patterns of aflatoxicol and aflatoxins B_1 and M_1 in blood and milk of cows given aflatoxin B_1 *Am. J. Vet. Res.*, **44**, 1753–1756.

Trucksess, M.W., Stoloff, L., Young, K., Wyatt, R.D. and Miller, B.L. (1983b) Aflatoxicol and aflatoxins B_1 and M_1 in eggs and tissues of laying hens consuming aflatoxin-contaminated feed. *Poult. Sci.*, **62**, 2176–2181.

Van Egmond, H.P. (1989) Current situation on regulations for mycotoxins. Overview of tolerances and status of standard methods of sampling and analysis. *Food Addit. Contam.*, **6**, 139–188.

Van Rensburg, S.J. (1977) Role of epidemiology in the elucidation of mycotoxin health risks. In *Mycotoxins in Human and Animal Health*. Eds J.V. Rodricks, C.W. Hesseltine, and M.A. Mehlman, pp. 699–711. Pathotox Publishers, Park Forest South, IL.

Vesonder, R.F., Ciegler, A. and Jensen, A. (1973) Isolation of the emetic principle from *Fusarium* infected corn. *Appl. Microbiol.*, **26**, 1008–1110.

Wang, T.V. and Cerutti, P.A. (1980) Spontaneous reactions of aflatoxin B_1 modified deoxyribonucleic acid *in vitro*. *Biochemistry*, **19**, 1692–1698.

Wogan, G.N. (1969) Metabolism and biochemical effects of aflatoxins. In *Aflatoxin*. Ed. L.A. Goldblatt, pp. 151–186. Academic Press, New York.

Wong, J.J. and Hsieh, D.D. (1976) Mutagenicity of aflatoxins related to their metabolism and carcinogenic potential. *Proc. Nat. Acad. Sci. USA*, **73**, 2241–2244.

Wu, M.T., Ayres, J.C. and Koehler, P.E. (1974a) Production of citrinin by *Penicillium viridicatum* on country cured ham. *Appl. Microbiol.*, **27**, 427–428.

Wu, M.T., Ayres, J.C. and Koehler, P.E. (1974b) Toxinogenic aspergilli and penicillia isolated from aged, cured meats. *Appl. Microbiol.*, 1094–1096.

Yoshizawa, T., Mirocha, C.J., Behrens, J.C. and Swanson, S.P. (1981) Metabolic fate of T-2 toxin in lactating cow. *Food Cosmet. Toxicol.*, **19**, 31–39.

Yoshizawa, T., Sakamoto, T., Ayano, Y. and Mirocha, C.J. (1982) 3'-Hydroxy T-2 and 3'-hydroxy HT-2 toxins: New metabolites of T-2 toxin, a trichothecene mycotoxin in animals. *Agric. Biol. Chem.*, **46**, 2613–2615.

10 Fermented meat production and consumption in the European Union

S. FISHER and M. PALMER

10.1 Introduction

Fermented meat products are just one category of processed meat. Because of the diversity and impreciseness of the information available on the processed meat market, it is extremely difficult to make a reliable estimate of the size of the fermented meat sector. Fermented products are commonly defined as those which have been subject to the action of microorganisms or enzymes so the desirable biochemical changes cause significant modification to the basic product. However, from the products included in the data on the processed meat market, it is not possible to reliably identify those which may have undergone this process.

This chapter analyses the processed meat market, and to focus more specifically on the fermented meat market, in the context of the total market for meat in the European Union (EU). The size of the EU market is put into perspective by comparing trends in production and consumption with that which has occurred in the rest of the world over the last twenty years.

10.2 Trends in world meat production and consumption

World meat production increased significantly throughout the 1970s and 1980s. Figures for 1989, published by the Food and Agriculture Organization of the United Nations (FAO), showed an increase of more than 50% over the two decades, to a total of 169 million tonnes carcass weight equivalent (Table 10.1). This is a dramatic increase for a basic food commodity. It does, however, equate with the estimated rise in population over this period.

Table 10.1 World meat production ($\times 10^3$ tonnes, carcass weight) (source: FAO estimates)

Meat	Average 1969–71	Average 1979–81	1989
Beef and veal	38 871	44 090	49 436
Mutton and lamb, incl. goat meat	7 321	7 516	8 838
Pig meat	38 619	51 699	67 460
Poultry meat	17 760	26 518	37 817

The rate at which production of the edible products of cattle, sheep, goats, pigs and poultry has expanded varied considerably. Beef and veal production rose by 27% between 1969 and 1989, mutton, lamb and goat meat by 20%, pig meat by 75% and poultry meat by 113%.

Clearly, the largest expansion has taken place in the intensive livestock sectors. This has been partly because of their shorter production cycle and partly because of their increasing price competitiveness. The latter has come about by the industrialisation of the production process in many parts of the world.

Interestingly, whereas at the beginning of the 1970s beef production and pig meat production were more or less equal, two decades later pig meat production had overtaken beef to account for 40 per cent of the world's meat output. Since pig meat is not eaten by Muslims, pigs are scarce in North Africa and in large parts of Asia. However, they have always been important in Europe and in East and South-East Asia. By the end of the 1980s, Asia had overtaken Europe as the world's largest pig meat producing region. Most of the increase in this region was accounted for by China, rising from 9.64 million tonnes in 1969–71 to an estimated 22 million tonnes in 1989. Within Europe there was a significant increase in pig meat production throughout the 1970s and 1980s in the EC. Since its formation, the EU's reliance on imported meats has fallen dramatically.

In the twenty years to 1989, central and North America maintained its position as the world's largest beef producing area. Europe continued to rank second but beef production in the EU was overtaken by that in the former USSR. A feature of world beef production is the disparity between cattle populations and meat output. Although the developing countries of the world have two-thirds of all cattle, their beef production only accounts for 30 per cent of total output. There are two explanations for this, firstly, the differences in meat yields because of more feeding, and secondly the fact that in the developing countries cattle are kept to a greater extent for traction and milk.

Poultry farming has become the most rapidly expanding type of livestock farming, owing to rapid growth and feed efficiency. As opposed to cattle and pigs, poultry are more evenly distributed throughout the world since they are eaten by all peoples and are traditionally kept for egg production. Production in developing countries increased relatively rapidly throughout the 1970s and 1980s from 5 million to 13 million tonnes, while production in the industrialised world rose from 12.5 million to 24.7 million tonnes.

Having risen at a much slower rate than the other meats, world production of sheep and goat meat still only accounts for a small proportion of total meat output. Interestingly, whereas Oceania was the world's second largest region of production after Asia at the beginning of the 1970s, by the end of the 1980s production had fallen below that of Europe.

Table 10.2 Estimated per capita meat consumption in selected areas of the world (kg) (1989) [Source: MLC, USDA, SOEC)

Area	Beef and veal	Sheep and goat meat	Pig meat
USA	45	1	29
Canada	40	–	33
Argentina	64	3	–
Uruguay	56	24	–
EC-12	23	4	40
Eastern Europe	16	1	45
China	1	1	19
Japan	9	1	18
Australia	39	22	2
New Zealand	35	30	–
USSR	30	3	24

The amount of meat consumed per head of population differs widely across the world, as does the meats preferred (Table 10.2). For example, while beef is the preferred meat in South America, pig meat is the largest element of meat consumption in eastern Europe and in China. In Japan and South Korea poultry is preferred. Table 10.2 illustrates the preferences for the different types of meat in selected parts of the world.

It is not surprising that in developing countries the main factors which have led to an increase in meat consumption are rising incomes and expanding populations. In developed countries these factors are becoming less important. Consumers' concern for diet and health, and even animal welfare, are now significant.

10.3 Trends in EU meat production and consumption

Although at the end of the 1980s the world's largest single producer of meat was the USA, followed by China, as a group the EU was the largest meat producing area. It accounted for almost 20% of world production. Whereas in the case of beef, veal and poultry meat, the EU was second to the USA, EU pig meat production was only surpassed by that of China. The EU was however the world's largest producer of sheep and goat meat

Table 10.3 Meat production ($\times 10^3$ tonnes, carcass weight) in the EC and other selected areas of the world (1989) (source: FAO estimates)

	Total*	Beef and veal	Mutton and lamb†	Pig meat	Poultry meat
EC-12	28 530	7 515	1 063	13 207	5 950
USA	28 344	10 655	157	7 176	10 088
China	27 178	662	821	22 070	2 767

* Includes buffalo meat and horse meat.
† Includes goat meat.

Table 10.4 Per capita meat consumption (kg) (1989) in the European Community (source: SOEC)

Country	Beef and veal	Sheep and goat meat	Pig meat	Poultry meat	Total*
Belgium/Luxembourg	20.8	1.8	46.7	15.8	97.9
Denmark	19.1	0.8	64.7	11.7	105.6
Germany†	22.8	1.0	58.7	11.4	99.0
Greece	23.7	14.6	23.3	16.0	84.0
Spain	11.7	5.9	45.1	22.5	94.0
France	30.4	4.9	37.4	21.1	109.8
Irish Republic	19.1	7.1	35.6	19.9	88.8
Italy	26.6	1.8	30.9	19.6	87.8
The Netherlands	19.3	0.7	46.5	16.7	87.4
Portugal	14.1	3.2	27.6	17.7	70.6
UK	19.2	7.2	24.2	18.9	73.8
EC-12	22.6	3.8	39.7	17.6	92.5

* Includes horse meat, rabbit, game, etc.
† Data based on former West Germany.

Average annual per capita consumption of meat in the EU, at 92 kg, falls just below that of the USA and Canada. However, there is a marked difference in consumption between both the individual member states and the different species (Table 10.4).

Total meat consumption is highest in France (110 kg), ranking first in beef and veal consumption and second in poultry meat consumption. In comparison consumption is least in Portugal (71 kg). Denmark and Germany (based on FRG data) are the second and third largest meat consumers because of their high intake of pig meat. Pig meat includes bacon, and like all other meats consumption figures includes that of processed meats.

A characteristic of societies which are comparably affluent and have a higher urban population is the demand for a higher proportion of processed meat products. There has been a dramatic increase in world production and consumption of processed meat products over the last two decades. Much of the growth has occurred in the EU. It is to this market that the remainder of the chapter is devoted.

10.4   Production of processed meat in the European Union

Having identified earlier that overall EU meat consumption rose significantly throughout the 1970s and 1980s, and particularly pig meat consumption, questions are raised about in what form the meat is consumed.

In all member states, pig meat is the most widely eaten meat, and in many cases a significant proportion of this is consumed as processed meat products. Beef and poultry are also consumed in a processed form but to a lesser degree than pig meat, while lamb is the least processed meat.

EU meat consumption figures published by organisations such as SOEC, OECD and FAO do not make a distinction between 'fresh' meat and processed meat products. Therefore data used to estimate the size of the processed meat market need to be derived directly from various national sources. As a result a country-by-country comparison is not feasible because of the different definitions that exist for certain products and secondly the presence of national 'specialities'.

As implied in the Introduction, fermented meats are those that have been subjected to a combination of chemical curing, microbiological fermentation and drying which then makes them safe to eat up to several months after production. Many fermented products are concentrated in the delicatessen area. Those that are generally recognised as possibly being fermented include German salami, Italian salami, *cervelat, peperoni, chorizo, mortadello, Lebanon Bologna, merguez, Kochsalami, Thueringer, jerky, nham, Teewurst, longaniza,* farmers sausage and country ham.

Because of the difficulty in extracting accurate figures on the production and/or consumption of fermented products from total processed meat products, it is only possible to make estimates of the size of the fermented meat market. Similarly up to date data does not exist. Given the differing sources in each country it is only possible to construct a benchmark around a common point for which there is most recent data. Specific reference is made to dried sausage and dried meat, not all of which will be fermented meats.

Many sausage products are of either German, Italian or Spanish origin. In fact, these three countries account for approximately 80% of estimated EU production of fermented sausages (see Table 10.5). Specific figures on

Table 10.5 EU production ($\times 10^3$ tonnes) of fermented sausages (1988) (source: Reading University, from various national statistics)

Country	Raw dry sausages		Approximate total
	Smoked	Unsmoked	
Belgium	10.8	1.2	12.0
Germany*	168.0	28.0	280.0†
Denmark	2.0	–	2.0
Spain	26.0	104.0	130.0
France	4.8	90.2	95.0
UK	<0.1	<0.1	<0.1
Greece	9.0	–	9.0
Italy	28.2	112.8	141.0
Irish Republic	<0.1	<0.1	<0.1
The Netherlands	>16.0	not applicable	20.0
Total			689.0

* Data based on West Germany.
† Approximately 30% of output is in the form of spreadable raw sausage.

either production or consumption of many of the fermented meat products are not available, partly because a large proportion of output is generated by small businesses. In Italy, for example, around one-half of production is generated by non-industrial concerns.

10.4.1 Germany*

The German processed meat market is the largest in the EU. The 288 factories employing 20 or more personnel produced 1.05 million tonnes in 1989. This excludes frozen ready meals and meat conserves but represents a 12% rise compared with 10 years earlier. However, this does not include the thousands of products produced by the small factories and from an estimated 25 000 individual butchers.

Sausages.   Sausage production as a proportion of total production has continued to account for roughly 75%. Over 1500 different variations of sausage, based on varying combinations of mainly pork and beef, are produced by both large manufacturers and small butchers. They fall into three main categories; *Rohwurst, Bruhwurst* and *Kochwurst*. However, only Rohwurst can be categorised as fermented sausage.

Rohwurst.   This is made from spiced raw meat (beef, pork or bacon). The meat is cured and dried by a special process of smoking and/or air-drying. Most Rohwurst are sliced before eating although some are soft enough to be used as a spread. Two examples of this type of German sausage are salami and cervelat.

Salami is the generic term for a dry-smoked, uncooked sausage, usually medium chopped and medium seasoning. Fermentation takes place at 20–32°C under humidity of 85–95% for 18–48 hours if a starter culture is added, or 5–9 days if not. It is then usually hot-smoked to an internal temperature of 55–63°C, then dried slowly at 15–24°C for several weeks or months. The following varieties are examples of salami: *Plockwurst* (three parts beef to one part pork, smoked and suitable for slicing); *Landjäger* (60% beef, 40% pork, flavoured with caraway seeds and garlic, cured, cold-smoked and air-dried, square in section for slicing); *Teewurst* (best quality finely minced pork and beef, usually highly spiced and suitable for spreading); and *Mettwurst* (regional differences, but basically a mixture of pork and beef and seasoned with sugar, paprika and nutmeg; it can vary from spreadable to very hard).

Cervelat are semi-dry, smoked sausage, usually finely chopped and mild-to-medium strength seasoning. Fermentation takes place at the same temperature and humidity as salami for 5–7 days. It is then smoked at 35–

---

* Data are for the former West Germany.

50°C for 12–48 hours before being dried slowly for 10–25 days. Varieties of cervelat are *Bierwurst, Rohe Krakauer, Schlackwurst* and *Schmierwurst*.

Most of the German smoked sausage producers are located in the Westphalia area which has an enviable reputation for quality sausages.

Bruhwurst.   These are made from finely minced meat, usually a mixture of beef or pork with bacon or bacon fat added, together with spices and seasoning. They are scalded by the manufacturer after undergoing a quick smoking process. Bruhwurst should be steamed, boiled or heated in the can before eating.

Kochwurst.   These are cooked sausages having been scalded, boiled and steamed. The different varieties of liver sausage (*Leberwurst*) and black pudding (*Blutwurst*) come under this category. Many Kochwurst contain large chunks of meat. They are usually eaten cold, either sliced or as a spread.

Other meats.   Processing factories produced nearly 150 000 tonnes of ham and smoked *Speck* in 1989 (Table 10.6). Probably about two-thirds of this can be categorised as fermented meat. Again production is concentrated in Westphalia. The area has given its name to one of the most popular hams. The *Westphalia ham* is produced by natural curing with rubbed salt and brine. After 3–4 weeks, the hams are smoked over resin-free beech and ashwood fires. This process may take up to five weeks.

Annual consumption of meat products amounted to 1.7 million tonnes by the end of the 1980s, representing about 42% of total meat consumption. This translated into a figure of 28.2 kg per head per year. Overall, sausages accounted for about 70% of all processed meat consumed. There has been relatively little change in the makeup of sausage consumption over the last twenty years, with Bruhwurst remaining the most popular. Consumption of fermented sausages — Rohwurst — amounted to 5 kg per head per year in 1989. The consumption of other potentially partially fermented meats — *Schinken* and *Speck* — also amounted to approx-

Table 10.6 Production of meat products ($\times 10^3$ tonnes) in Germany* (1989) (source: Bundesverband der Deutschen Fleischwarenindustrie)

Sausage products	768.9
Rohwurst	258.5
Bruhwurst	352.8
Kochwurst	157.6
Ham smoked speck	146.8
Other	137.8
Total	1 053.5

* Data refer to the former West Germany.

imately 5 kg per head per year. In total, therefore, the consumption of fermented meat and meat products amounted to an estimated 600 000 tonnes.

10.4.2 France

As in Germany, the French meat processing industry is composed of both the industrial sector and production by traditional butchers. There has been dramatic concentration in the industrial sector over the last two decades, with the number of firms falling to just over 400 by the end of the 1980s. However, output continued to increase to approximately 865 000 tonnes by 1989. At the same time, it was estimated that the 24 600 pork butchers accounted for 20% of processed meat production. In total, therefore, the production of processed meats exceeds over 1 million tonnes, roughly equivalent to the German industrial sector. Pig meat accounts for approximately 94% of the raw materials used in the manufacture of French meat products. Table 10.7 provides a break down of industrial production by the different types of product.

Dried sausages (*saucisson sec*) and dried hams (*jambon seches fumes*) can be identified as possibly being fermented products. Together they account for just 18% of the total. However, other varieties of sausage classified as being fermented, e.g. *mortadello*, are categorised as cooked sausage. Total fermented meat production is therefore possibly in excess of 160 000 tonnes.

Table 10.7 Production of meat products ($\times 10^3$ tonnes) in France (1989) (source: Fédération de la Charcuterie, Salaison, Filière Viande)

Salaisons	335.5	+3.7*
Raw, salted meats, including hams	25.1	+11.6
Dried, smoked meats, including hams	70.5	−2.6
Cooked meats, including hams	239.9	+4.9
Dried sausages	87.9	−1.8
Charcuterie	355.9	+3.2
Pâtés, terrines, rillettes	120.0	+2.0
Products made from head meat	23.5	+7.8
Products in pastry	13.2	−5.1
Cooked sausages	52.8	−5.8
Fresh sausages	126.5	+8.1
Other	19.8	+8.4
Beef-based products	23.0	+49.5
Ready-cooked dishes	62.2	+7.7
Total all products (includ. lard, fats)	865.4	+4.0

* Column shows change (%) 1989/1988 for each type of product

Sausages. There are many varieties of *saucisson sec*, many with regional characteristics, for example: *saucisson d'Arles* (a mixture of coarsely chopped pork and finely minced beef, seasoned with garlic, pepper and red wine, air dried up to six months), *saucisson de Metz* (70% lean beef, 10% pork, 20% bacon, seasoned with pepper and coriander, air-dried for five days and cool smoked for three days), *Rosette* (coarse-chopped pork), *Manage* (pork) and *Montagne* (coarse-chopped pork, very thin sausage).

Other meats. Approximately 70 000 tonnes of dried, smoked meats are produced by the manufacturing sector each year. Slightly more than half of this is accounted for by ham. Again the varieties of ham have regional characteristics, for example *jambon blanc* (similar to York ham) and *jambon de Paris*.

Demand for processed meat products is unable to be met by domestic production and consequently there is a high import requirement. In the case of dried sausages, in 1989 imports exceeded exports by 4 300 tonnes. Consumption is therefore estimated at in excess of 92 000 tonnes if that produced by traditional pork butchers is included. Denmark is the main source of French sausage imports.

Net imports of dried meats totalled 15 200 tonnes in 1989, resulting in total domestic consumption in the region of 100 000 tonnes. Trade statistics suggest that rougly 5000 tonnes of Parma ham is consumed per annum.

Overall then, if all dried sausages and meats were to be fermented, annual consumption of fermented meat and meat products is estimated to be at least 190 000 tonnes, equal to approxiamately one-third of that in Germany.

10.4.3 Italy

The Italian meat processing industry is extremely fragmented. There are in excess of 2000 producers but in 1988 only 330 employed over 20 personnel. For many meat products, the four largest companies account for only 10% of total production. Contrary to trends in other food industries there was a significant increase in the number of small processed meat producers during the 1980s. Small artisan producers account for roughly 40% of total production.

Nearly 60% of all pig meat consumed in Italy is in the form of processed products. Consumption has remained relatively unchanged throughout the 1980s, totalling 947 000 tonnes in 1989. Of this, approximately 30% can be estimated as being fermented meats. Most of that which is consumed is domestically produced, with total imports of processed meat being in the region of 50 000 tonnes. Relatively little is exported, and then mostly as Parma ham.

Table 10.8 Consumption of processed meat ($\times 10^3$ tonnes) products in Italy (source: Marché International des Viandes Transformées)

Product	1985	1989
Uncooked ham	139	142
Cooked ham	279	295
Salami	147	138
Mortadella	198	195
Sausages	32	37
Other	145	140
Total	940	947

Sausages. Sausages account for the largest proportion of fermented meat consumption in Italy, with the absolute level having changed very little in recent years (Table 10.8). Most Italian premises have their own speciality of sausage but the most widely known is *mortadella* or Bologna sausage. There are many recipes for mortadella although it is calaimed that only those with are pure pork are the real sausage. They were originally made from pigs fed on chestnuts and acorns. The sausage is allowed to ferment at 30–38°C for 2–7 days then it is cooked to an internal temperature of 65–70°C before cooling. Mortadella is air-dried during maturation.

Other sausage varieties include *cotechino* (lean and fat pork, flavoured with white wine and subtly spiced), *peperone* (coarsely chopped 40 pork, 60% beef and highly seasoned with red peppers, fennel and spice) and *Genoa* (extra lean pork and spices, air dried).

Salami is categorised separately because of the many varieties. It is the only true fermented pig meat product. Many salami are pure pork whereas others are a mixture of pork and beef, varying by region. Some salami are pickled in brine and then air-dried while others are smoked. Varieties of salami include *Calebrese* (contains pieces of white fat) and *Milanese* (at least 50% lean pork and 20% pork fat, seasoned with garlic, white wine, peppercorns and sugar, air-dried). In general, salami is more highly spiced the further south you go. Production of salami has remained fairly static in recent years at around 102 000 tonnes.

Other meats. Consumption of uncooked ham, or dried ham, which undergoes a partial fermentation process, at 142 000 tonnes equates to between 22 and 23 million pieces. Production is concentrated in three regions — the area south of Parma, the region of San Daniele and the province of Veneto. The production of *Parma ham*, by far the most popular, amounted to 7.5 million pieces in 1989, approximately 46 000 tonnes. The number of firms engaged in its production totalled 215. This represents several years of decline. The production of *San Daniele ham* is much lower, at less than 2 million pieces.

Although Parma ham is world famous, Italian exports were still only slightly less than 6000 tonnes in 1989. The main recipient is France.

10.4.4 Spain

After Germany, Spain has the second highest per capita consumption of processed meat products in the EU. The market is dominated by pig meat products, especially sausages.

In 1989, there were 3500 meat processing plants, producing approximately 700 000 tonnes of pork-based products. There are no data available on the quantity produced by small units and butchers.

Production in Spain is divided into two main areas, cooked production and cured production which includes air-dried products. Historically, the volume of cured production has been significantly higher than that of cooked, but by 1988 the gap between the two had closed and in 1989 cured products accounted for less than half.

Dried ham accounts for the largest proportion of processed meat production — 140 000 tonnes — but production of *chorizo*, sausages, salami and other dried products is not insignificant. Together with dried ham they accounted for 41% of total production, or 290 000 tonnes in 1989.

Sausages. *Chorizo* is the most popular dried sausage, with production estimated at 62 000 tonnes. It is a highly spiced sausage made from coarsely cut pork with added peppers. It is fermented at 30–38°C for 2–6 days and then dried for several weeks or months at 20–25°C. *Salchichan* is made in a similar way to *chorizo* but is less highly spiced. *Fuet* are long thin *salchichan*, the ends of which are often tied together like a horse shoe. Production of *salchichan* and *fuet* is estimated at 47 000 tonnes. Other cured sausage products include salami, *mortadella*, air-dried *longaniza* and *morcialla*. Production is estimated at 11 000 tonnes.

Other meats. Dried ham accounts for approximately 45% of cured meat production. It is a very traditional market with a large number of regional hams being produced. The most popular is *Serrano ham*. Production in 1989 is estimated at 125 000 tonnes. *Iberico ham* is also well known for its quality. However this ham is produced in an area where African swine fever is still a problem and so exports are impossible. Production is around 35 000 tonnes a year.

Two major events have occurred in the industry since the mid-1980s and which have increased the rate of development. Firstly, Spain's accession to the EC in 1986 has forced the pig meat industry, especially, to adapt rapidly to competition with other member states. Secondly, it was not until

January 1989 that most of the Spanish territory was declared free from African swine fever and May 1989 before many areas were declared free from classical swine fever. Until then, Spanish pork products could not be traded within the Union.

Prior to accession, Spain was a net exporter of processed meat products with few imports of any category. On accession, because of health reasons, exports of pig meat from most of Spain to other member states were banned. However, 1990 trade data shows that producers have taken advantage of the country being declared officially swine fever-free. Exports of dried, and therefore potentially fermented, pig meat products to other member states amounted to 830 tonnes. Nearly two-thirds of this was in the form of sausage products. In comparison, imports of dried pig meat products totalled 340 tonnes.

10.4.5 Belgium

Compared with two of its neighbouring countries, France and Germany, the Belgian processed meat market is relatively small (Table 10.9). In 1989, total production was estimated at 247 000 tonnes. Approximately 40% of all production is destined for export, primarily to France. There are, however, some product areas where production cannot meet domestic demand. In 1989, 20 000 tonnes of processed meats were imported. In the case of fermented meats, however, Belgium is a significant net exporter.

Fermented meat production is split fairly evenly between sausages and other meats. However, in terms of consumption, dried sausages are far more popular. They account for roughly 9% of total processed meat consumption. Cooked hams rank the highest in popularity, accounting for 34 000 tonnes (3.3 kg per head) or 22% of total consumption.

Sausages.   The market for dried sausages remained fairly static during the late 1980s. In terms of total imported supplies, there has been little fluctuation since 1987 but France increased its share of the market at the expense of Italy. This is because of the Italians placing more importance on the development of export outlets for Parma ham. There are several varieties of domestically produced sausages. They are usually medium

Table 10.9 Estimated fermented meat supplies (tonnes) in Belgium (1989) (source: Marché International des Viandes Transformées and MLC estimates based on customs data)

Meat	Production	Imports	Exports	Consumption
Dried sausage	18 500	1975	4975	15 500
Dried and smoked meats	17 150	1491	9641	9000

chopped and a mixture of pork and beef. *Ardenner, Courant, Edelkost, Fijnkost* and *Ole*. There is also a variety of horse meat sausage, *Boulogna*.

Other meats.   The production of dried and smoked meat is dominated by dried hams. The most popular variety is *Ardennes ham*. Approximately one-third of production is destined for export to France. Although imports of dried meats fell during the late 1980s, competition between French, Italian and German supplies is intensified. Parma ham is extremely popular with the Belgian consumer, as too is Westphalia Ham. They accounted for 47 and 37% of total imports in 1989, respectively.

10.4.6   Denmark

Denmark's meat processing industry is primarily geared towards export. In fact Denmark is the world's largest exporter of processed meats. The industry concentrates on producing a relatively narrow range of standard-quality, reasonably priced pork products such as bacon, canned hams and shoulders, luncheon meat, frankfurters and salami. Exports have been in the region of 200 000 tonnes since the mid-1980s. Canned meats account for 80% of exports. The remainder, 35 000 tonnes includes Danish salami.

Sausages.   Danish salami is a mixture of finely chopped pork, beef, veal and pork fat, sometimes containing cereal. It is usually bright red in colour. Examples of varieties of salami are *Sondertysk spegepolse* and *spegepolse*. Annual production of salami amounts to around 26 000 tonnes, approximately half of which is estimated to be fermented.

10.4.7   The Netherlands

In 1990 there were 160 locations for the processing of meat and meat preserves in The Netherlands. In 44 of these, annual production exceeded 1000 tonnes. Total production in 1990 amounted to 243 000 tonnes, excluding bacon, an amount which has varied little since the mid-1980s. However, this figure does not include the products produced by individual butchers. They are estimated at 30 000 tonnes. Sausages account for the largest proportion of total processed meat production, followed by hams.

The Netherlands produce well in excess of their domestic requirements and as a result exports of processed meats amounted to 84 000 tonnes in 1990. The main export commodities are luncheon meat and cooked ham.

Table 10.10 suggests the production of fermented meats in The Netherlands accounts for a very small proportion of total output.

Sausages.   Fermented sausage production is estimated to account for roughly one-third of total sausage production. Some of the most well

Table 10.10 Output of selected processed meats ($\times 10^3$ tonnes) in The Netherlands in (1990) (source: PVV)

Sausages	90.4	Ham	57.3
dry	13.4	raw	3.3
cooked	18.1	cooked	34.5
smoked	18.4	shoulder	19.5
frankfurter	20.5	Luncheon meat	34.6
liver	17.9	Beef products	8.3
other	2.1	Liver products	9.1

known types are *Theeworst*, a semi-dry smoked sausage, soft in texture for spreading, *Snijworst*, a variety of cervelat and *Gelderse rookworst* a moist, course-chopped beef sausage, heavily smoked.

Other meats.   Cooked ham dominates the production of pork. Luncheon meat production is of a similar magnitude. As a result, very little is produced in the way of fermented meats.

10.4.8   United Kingdom

Approximately 20% of all meat consumed in the home is estimated to be in the form of processed meats (excluding bacon). The proportion has changed very little over the last decade. Pork continues to account for around 46% of the total consumed but the largest area of growth has been in the poultry sector. In 1990, total household processed meat consumption was estimated at 640 000 tonnes, carcass weight equivalent. This equates to roughly 10 kg per person per year.

In many product areas, demand cannot be met by domestic output. One such sector is the continental-style products. Consumption of these products increased considerably in the late 1970s and early 1980s. In more recent years, the market has stabilised and household consumption in 1989 was estimated at 26 000 tonnes product weight. Of this, around 40% is accounted for by pâté and 60% by continental sausages.

Sausages.   There are two basic types of continental sausage. The first group is cooked sausage which includes liver sausage, garlic sausage, frankfurters, etc. The second group is the dried, fermented sausage which includes salami. As the Table 10.11 illustrates, those that can be identified as possibly being fermented sausages are much less popular than the cooked sausages. Also, the vast majority of that which is consumed is imported. In 1990, the UK imported 4700 tonnes of dry sausage, mainly from Denmark and Germany. The UK is one of only a few European countries not to have these sausages as an indigenous product or therefore to have developed the technology of fermenting and drying.

Table 10.11 Estimated UK household consumption (tonnes) of continental sausages (1989) (source: MLC based on National Food Survey Data)

German salami	260	Mild garlic	260
Salami	1040	Sliced garlic	260
Pepperoni	260	Garlic	2340
Pepperami	260	German liver	260
Kabanos	520	Liver	1040
Kalsinos	250	Saveloy	520
Pork breakfast	780	Frankfurters	1820
Ham	520	Polony	2080
Smoked	520	Bratwurst	780
French	260	Bierwurst	260
Continental	780	Herta German	260
French garlic	1040	**Total**	**16 380**

Other meats. Because of the lack of facilities for fermenting, the consumption of other fermented meats is nearly all of foreign origin. The Parma ham market, for example, although growing is still very small.

10.4.9 Other EU countries

Together with the UK, production of fermented meat in Greece, the Irish Republic and Portugal is relatively insignificant. There is also a lack of information on the market in these countries.

10.5 Conclusion

Because of the diversity of sources from which information on the fermented meat market has been collected, it is very difficult to try and draw a comparison of production and consumption in each of the EU member states. Suffice to say, that the two major producing and consuming countries are Germany and France, followed by Italy and Spain.

To some extent, countries specialise in specific product areas and in many cases produce more than is demanded on the domestic market. Consequently, there is a significant internal Union trade (Table 10.12). The level of trade with non-EU countries is much less. Information on specific fermented meat products cannot be extracted from published trade data. Fermented sausages, however, were included in the tariff code '1601.00–91: Uncooked Sausages of Meat, Offal or Blood, Dry or for Spreading, Excluding Liver'.

In 1990, figures showed that almost 30 000 tonnes of uncooked sausage was traded within the that year. As noted earlier in this chapter, France has a high import requirement for processed meat products despite being one of the largest producers. In 1990 its imports of dried sausage from other EU countries far exceeded its exports and accounted for one-third of total EU

Table 10.12 Intra-European Union trade (tonnes) in uncooked sausages (1990) (source: Nimex)

From:	EU	Belgium	Denmark	Germany	Greece	Spain	France	Ireland	Italy	Netherlands	Portugal	UK
To:												
EC	29 853	5 604	9 675	5 155	131	534	3 790	39	4 663	246	0	16
Belgium	2 743	0	31	592	0	67	1 321	0	668	64	0	0
Denmark	273	2	0	250	0	0	8	0	12	0	0	1
Germany	5 686	1 079	226	0	96	68	2 346	17	1 785	67	0	2
Greece	243	10	179	10	0	0	0	0	44	0	0	0
Spain	146	0	26	60	6	0	5	1	52	2	0	3
France	10 322	487	7 118	1 119	0	318	0	0	1 271	27	0	0
Ireland	64	2	1	20	0	0	3	0	4	4	0	7
Italy	377	0	4	353	0	5	11	0	0	4	0	0
Netherlands	5 317	3 882	5	1 278	0	16	40	0	93	1	0	3
Portugal	118	5	34	16	0	47	14	0	1	1	0	0
UK	4 564	137	2 051	1 457	29	13	42	21	733	81	0	0

imports. The other large importers were Germany, The Netherlands and the UK. In the case of The Netherlands and the UK, the volume of imports contrasts with an insignificant quantity of exports. However, Germany exports nearly as much dried sausage to other EU countries as it imports. This emphasises the supply and demand imbalance of the individual products. In addition to this, Germany imported approximately 4000 tonnes of dried sausage from non-EU countries, mainly Hungary.

It was previously mentioned that the Danish processing industry is primarily geared for export. Table 10.12 susbstantiates this, showing that Danish exports of dried sausage to other EU countries accounts for one-third of EU exports. In addition, it exported nearly 7000 tonnes to outside the Community. Since in most member states the production of other fermented meats appears to be less than that for fermented sausage, it is assumed that the level of intra-Union trade would be much less. Unfortunately trade in, for example, Parma ham cannot be readily extracted from customs data.

As far as future developments in EU production and therefore trade in fermented meat products are concerned, much will depend on the implementation of legislation in respect of the completion of the Single Market by 31 December 1995. Some relevant hygiene, compositional and labelling legislation is under review at the time of writing. Agreement in some areas is likely in the near future, but it remains difficult to determine the precise effect on Union output. The only definite outcome is that some member states will suffer from the removal of technical barriers to trade which in the past have protected national interests, giving opportunities to the industries elsewhere.

Index

The manufacturer's authorised representative in the EU is Springer
Nature Customer Service Centre GmbH, Europaplatz 3, 69115 Heidelberg,
Germany. If you have any concerns regarding our products, please
contact ProductSafety@springernature.com

Printed and bound by CPI Group (UK) Ltd, Croydon, CR0 4YY

23/04/2026
02095628-0001